Tsukemono

Ole G. Mouritsen · Klavs Styrbæk

Tsukemono

Decoding the Art and Science of Japanese Pickling

Photography
Jonas Drotner Mouritsen

Translation and adaptation to English
Mariela Johansen

 Springer

Ole G. Mouritsen
Department of Food Science
University of Copenhagen
Copenhagen, Denmark

Klavs Styrbæk
STYRBÆKS
Odense N, Denmark

Translation and adaptation to English by Mariela Johansen

Photography by Jonas Drotner Mouritsen

ISBN 978-3-030-57864-0   ISBN 978-3-030-57862-6   (eBook)
https://doi.org/10.1007/978-3-030-57862-6

This Springer imprint is published by the registered company Springer Nature Switzerland AG
The registered company address is: Gewerbestrasse 11, 6330 Cham, Switzerland

Contents

Preface

For the past five years or so Ole and Klavs have been preoccupied with exploring various aspects of the texture of food, referred to more formally as its *mouthfeel*. This quality has an incredibly important effect on the sensory experience of eating—whether a particular food is considered interesting, whether its 'taste' lives up to expectations, and whether people like or reject it. Another special focus of ours has been the fifth taste, umami, described as 'meaty,' 'savoury,' or 'brothy,' which is also central to our enjoyment of, and craving for, certain types of food. The end result of our work on these two topics has been the publication of a book devoted to each subject. Additional research has been carried out under the auspices of *Taste for Life*, an interdisciplinary Danish centre that combines the study of the scientific underpinnings of taste with wide-ranging efforts to disseminate this information to the general public—from school children to seniors—to emphasize how healthy food choices can be made more appealing and lead to a better quality of life.

These two factors relate particularly closely to the consumption of vegetables, which are a vital component of our diet. Nevertheless, many people will readily admit that on a daily basis they frequently fail to eat them in sufficient quantities. We think that the underlying reason is that there is little emphasis on preparing them in the right way. The secret often lies in ensuring that they remain crispy and crunchy, have visual appeal, and taste of umami.

These ideas prompted us to turn our attention to *tsukemono*—pickled foods made from vegetables and occasionally fruits that are preserved according to centuries-old traditions in Japan. Our point of departure was twofold. Klavs has extensive experience with preparing vegetables in high-end restaurants, at innovative gastronomic workshops, and in cooking schools. Ole has a lifelong passion for Japanese cuisine, coupled with a scientist's understanding of the chemical and physical principles that can influence the texture and taste of raw ingredients. This led us to a series of experiments in the kitchen and in the laboratory to test classical recipes for *tsukemono*, as well as to develop modern techniques for preparing these preserves. We make no secret of the fact that this venture was enlightening and has been a great deal of fun.

We decided to put our findings, along with the personal experiences we have accumulated over the years, together in book form in the hope of introducing both the art and the science behind these special pickled foods to a broad international audience. To do so, we are providing recipes and outlining techniques for preparing them at home using ingredients that are readily available from local sources or in stores that sell Asian products. But we also go well beyond simply explaining the secrets of making crisp *tsukemono*. We have included various aspects of the cultural history and traditions that are associated with this ancient culinary art. At the same time, we have unpacked some of the fascinating science that explains how the preservation methods work. We will furthermore describe their tastes and the healthful benefits and basic nutritional value to be found in the various types of pickles and show how easy it is to serve them on a daily basis to stimulate the appetite or as condiments to accompany vegetable, fish, and meat dishes.

Since the topic of our book is inspired by Japanese cuisine, we have found it natural and convenient to adopt Japanese expressions for some kinds of *tsukemono* and the various techniques used to prepare them. Also, there are sometimes no English terms that adequately convey their meaning. In cases where there are both English and Japanese terms for the same ingredient, for example, Chinese radish (*daikon*), we have used these terms interchangeably. We have included a glossary of these Japanese words with English explanations.

Our ultimate goal is to encourage the readers of this book to join us in a small culinary adventure that will allow us to expand and diversify our consumption of plant-based foods, which are so vital to our overall well-being. And along the way, there may be a few surprises. Most of us are familiar with the little mound of deliciously tangy pickled ginger that is served with sushi and sashimi, but how many have ever imagined that one could eat preserved cherry blossoms?

◘ Selection of *tsukemono*.

The People behind the Book

Ole G. Mouritsen

is a research scientist and professor of gastrophysics and culinary food innovation at Copenhagen University. His work focuses on basic sciences and their applications within the fields of biotechnology, biomedicine, and food. He is the recipient of numerous prizes for his work and for research communication. His extensive list of publications includes a number of monographs, several of them co-authored with Klavs Styrbæk, which integrate scientific insights with culinary perspectives and have been nominated three times for Gourmand Best in the World Awards. Currently, Ole is president of The Danish Gastronomical Academy and director of the National Danish Taste Centre *Taste for Life*, which is generously supported by the Nordea Foundation. This is a cross-disciplinary centre that aims to foster a better understanding of the fundamental nature of taste impressions and how we can use this knowledge to make much more informed and healthier food choices. Its extensive educational program reaches out to audiences of all ages, with a special effort directed toward children to shape their dietary habits from an early age. For many years, Ole has been fascinated with the Japanese culinary arts and in explaining the extent to which its techniques and taste elements can be adapted for the Western kitchen. In recognition of his efforts, he was appointed in 2016 as a Japanese Cuisine Goodwill Ambassador by the Japanese Ministry of Agriculture, Forestry, and Fisheries, and in 2017 the Japanese Emperor bestowed upon him The Order of the Rising Sun, Gold Rays with Neck Ribbon, *Kyokujitsu chujusho* 旭日中綬.

Klavs Styrbæk

is a professional chef who owns and operates STYRBÆKS together with his wife, Pia. By combining a high standard of craftsmanship, sparked by curiosity-driven enthusiasm, he has created a gourmet centre where people can enjoy excellent food and where they can come to learn and take their culinary skills to a whole new level. Klavs is especially committed to seeking out unique, local raw ingredients that can be incorporated into new taste adventures or used to revisit traditional Danish recipes that might otherwise be forgotten. This delicate balance between innovation and renewal is demonstrated in his award-winning cookbook *Mormors mad* (*Grandmother's Food*) (2006), which was honoured with a special jury prize at the Gourmand World Cookbook Awards in 2007. In 2008 and 2019 he was awarded an honours diploma for excellence in the culinary arts by the Danish Gastronomical Academy. Many of the recipes that appear in the books co-authored with Ole originated in the test kitchens at STYRBÆKS.

Jonas Drotner Mouritsen

is a graphic designer and owner of the design company Chromascope that specializes in graphic design, animation, and film production. His movie projects have won several international awards. In addition, he has been responsible for layout, photography, and design of several books about food, some of which have been nominated for Gourmand World Cookbook Awards.

Mariela Johansen

who has Danish roots, lives in Vancouver, Canada, and holds an MA in Humanities with a special interest in the ancient world. Working with Ole and Klavs, she has translated several monographs, adapting them for a wider English language readership. Two of these, *Umami: unlocking the secrets of the fifth taste* and *Mouthfeel: How Texture Makes Taste*, won a Gourmand World Cookbook Award for the best translation of a cookbook published in the USA in 2014 and 2017, respectively.

Thanks and Acknowledgements

In the course of our work on this book we have been helped and guided by a large number of colleagues and good contacts. Special thanks are due to:

- Our co-workers at *Taste for Life* for inspiration and collaboration regarding taste.
- Mathias Porsmose Clausen for providing unpublished microscopy images of the cell structures of various vegetables and *tsukemono*.
- The Japanese Embassy in Denmark for assistance in making contacts in Japan.
- Dr. Kumiko Ninomiya, Umami Information Center in Tokyo, for arranging visits to *tsukemono* factories in the Gunma Prefecture in Japan and for tracking down information about *tsukemono* production.
- Hideyo Shitara for his hospitality and for showing Ole around the Shitara *tsukemono* factory in Takasaki.
- Masami Kobayashi and Shuji Sawaguchi for showing Ole around the Shin-Shin *tsukemono* enterprise in Maebashi.
- Dr. Koji Kinoshita for priceless assistance with the correct use of Japanese words and expressions.
- Mette Holm for Japanese translation of the titles of the book chapters and for guidance regarding Japanese pronunciation.
- Anders Møller Pedersen from FSG Foods Scandinavia for providing a variety of Japanese specialty ingredients for the preparation of *tsukemono*.
- The Holistia Organic Market Garden in Odense for delivering top quality fresh produce.

The individuals who have generously made illustrations available for this volume are listed along with the picture credits.

Jonas Drotner Mouritsen has participated in the production of this book from its inception and has taken most the photographs.

This book was originally written and published in Danish, the mother tongue of the authors. The present volume is a fully updated and carefully revised version that was translated and adapted for a broader international audience by Mariela Johansen. Mariela undertook the challenging task of working with the interdisciplinary material to produce a coherent, scientifically sound, and very accessible book. This involved not only translating the text, but also checking facts, ensuring consistency, and suggesting new material and valuable revisions. The authors are extremely appreciative of her devotion to this project.

Tsukemono— a Japanese Culinary Art Based on the Science of Preservation

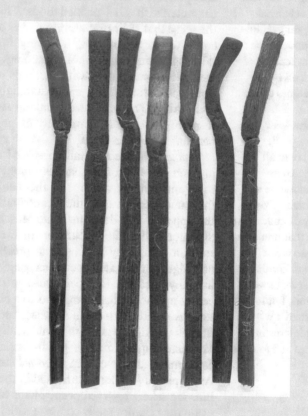

One of the best-kept secrets of Japanese cuisine, which the wider world has yet to discover in depth, is a range of side dishes known as *tsukemono* (つけもの, 漬物). The word, which is pronounced like 'tskay-moh-noh,' literally means 'something that has been steeped or marinated' (*tsuke* 'steeped' and *mono* 'things'). Most readers are already familiar with Japanese foods such as sushi, sashimi, ramen, teriyaki, tempura, and soy sauce. But where do *tsukemono* fit in? While they may not yet have appeared over the horizon in Western cuisines, these pickles are just as common a part of every traditional Japanese meal—breakfast, lunch, and dinner—as cooked rice and *miso* soup.

Tsukemono terminology. When *tsuke* follows another word it is changed to –*zuke*, for example in *miso-zuke*, which are vegetables marinated in *miso*. *Tsukemono* are sometimes called *oshinko, konomono,* and *okoko*—as *ko* means fragrant, the words can all be translated as 'things that have a good aroma.' These expressions underscore the idea that preparing *tsukemono* is not merely a method for preserving raw ingredients, but to an equal extent a way of creating interesting new tastes and aromas.

Although *tsukemono* are usually made from vegetables, some fruits, as well as a few rhizomes, and even flowers are also preserved this way. It is, therefore, more accurate to characterize them as 'pickled foods.' This is especially so in the sense that their preparation makes use of one or more conservation techniques, involving ingredients such as salt, sugar, vinegar, alcohol, seasonings, and fermentation media, in combination with methods including dehydration, marinating in salt and acidic liquids, fermentation, and curing. The process of making *tsukemono* amounts to more than just a simple way of preserving otherwise perishable fresh produce.

Tsukemono are normally prepared without any cooking and are eaten cold, so just about anyone can quite easily make many versions of them at home. Apart from their nutritional value, their contribution is to stimulate the appetite, add delicious taste sensations, and improve digestion, all while remaining an exceptionally elegant study in simplicity and esthetic presentation. The ordinary lightly marinated cucumber salad that is so common in many northern European cuisines can be tasty, but it comes up short when compared to a crisp Japanese *kyuri asa-zuke* made with small Japanese cucumbers. And even though pickled beets are wonderfully sweet and sour, the taste impressions we get from them are not nearly as complex as those from *shiro-uri kasu-zuke*, which are Asian pickling melons marinated in the lees left over from brewing sake. Preserved prune plums can be delicious, but again they cannot really compete with *umeboshi*, the small brined plum-like apricots that are placed in a mixture of plum wine and aromatic red *shiso* leaves.

When we eat *tsukemono* all of our five senses are engaged. The presentation of these pickles is simple and inspired by Zen esthetics—it combines different colours and shapes in a way that delights the eyes. When a piece of *tsukemono* moves past the lips and enters the mouth, the initial impression is a feeling of limpness, but the first bite reveals a completely different story. Its mouthfeel is surprisingly crisp and it is so crunchy that the sound immediately brings our sense of hearing into play. The taste impressions on the tongue are derived from salt and acid, as well as umami-rich compounds. Sometimes sugar and other seasonings that have been used to prepare the pickles enhance the flavour. Finally, when we chew on the *tsukemono*, the sense of smell is engaged as aromatic substances are released up into, and out through, the nose. These multi-sensory aspects of *tsukemono* are an intrinsic part of their appeal.

'The Taste and Smell of Home'

In traditional Japanese cuisine, *tsukemono* are the foods most closely associated with one's home. In his comprehensive work, *Japanese Cooking: A Simple Art*, the famous Japanese food writer Shizuo Tsuji states that for the Japanese these pickled foods are what bread and cheese are for the English and bread and wine for the French. The ways in which they are prepared are steeped in local traditions that are specific to communities, and even to individual families. These preparation methods are considered inviolable and dictate how the resulting *tsukemono* should taste.

Over the years, *tsukemono* have taken on a symbolic value and are regarded as a tangible expression of a mother's love for her family and concern for its well-being and good health. They are the very embodiment of solicitude and are an example of what is known as *ofukuro no aji* (an expression that describes the nurturing food prepared by one's mother). This is really no different from the nostalgia many of us feel for our own mother's home cooking and the taste of the comfort food of our childhood.

While *tsukemono* are evocative of 'the taste of home,' their special smell is even more strongly associated with a household. In a traditional Japanese house, this unmistakably pungent odour has permeated the whole building and greets one as soon as one opens the door. It is the heavy smell of *miso*, fermented rice bran, and the lees from sake brewing, or the sharp, acidic smell of vegetables in the pickling crocks. Although this is the so-called 'smell of home,' not everyone finds it particularly agreeable.

▪ A simple Japanese meal consisting of cooked white rice, *miso* soup, green tea, and a small assortment of *tsukemono*.

Tradition and Renewal

The population shift from the countryside to the urban centres that started during World War II has had a profound impact on many aspects of Japanese life, including its culinary culture. Globalization and the adoption of Western fast foods has led to a decrease in the consumption of *tsukemono* and, to an even greater extent, undermined the tradition of making them at home. Most of these pickled foods are now produced in factories and their popularity has been on the wane since 1991. Whereas they were once an important and integral part of a meal, many now think of them as optional. There has also been a movement away from the more traditional and stronger tasting varieties in favour of lightly marinated vegetables with a lower salt content and a less complex taste.

This development has taken place in parallel with another, more striking change in Japanese food culture, namely, that households no longer make *dashi* on a daily basis. It takes time to make *dashi* from scratch and time is a precious commodity in a modern society. *Dashi*, which is a broth made from seaweed (*konbu*) and a preserved fish product (*katsuobushi*), is ubiquitous in the Japanese kitchen. In fact, umami, the special savoury fifth taste, was originally identified by analysing *dashi*. It is used not only in soups but can also be regarded as the focal point around which all Japanese cuisine revolves. It is difficult to picture a real Japanese meal without cooked rice, *dashi*, and *tsuke-mono*, although the broth is now usually made from store-bought powders to which water is added.

The art of canning and pickling, which had virtually disappeared from many Western food cultures, is now enjoying a bit of a revival. This is also true of *tsukemono* in Japan, where there is a nascent, if somewhat nostalgic, movement to make them, once again, at home. And as has happened with other foods that have been mass-produced, the preferred taste of the *tsukemono* has undergone a change. The factory-made products with a variety of additives now on the market have a milder taste. As a result, many people find that those made according to the traditional preservation techniques have much stronger and more lingering tastes, which are quite different from those to which they have become accustomed. This is especially true of the intense taste and pungent odour of fermented products such as *kasu-zuke* and *miso-zuke*, which are made with sake lees and *miso* paste, respectively.

◘ Pickling and fermentation of vegetables in crocks and jars, stored in a cold, dark place.

An analogy can be made with one of the most traditional foods found in Nordic cuisine, marinated herring. In former times, the herring were prepared in the old-fashioned way—ungutted, whole fish were placed in a barrel with layers of salt and left for several months. The intestinal enzymes of the fish fermented and tenderized the flesh and it took on a strong, yeasty taste. Now they are usually made in factories that have introduced other ways of curing herring more cheaply and much more rapidly. The fish filets are marinated in a strong vinegar mixture, leaving the pieces firmer, but with a sharper, sour taste. Most consumers now prefer this taste, adopting it as the norm, and herring prepared in the traditional way have fallen out of favour.

Some Western food enthusiasts are once again turning to fermentation using natural lactic acid bacteria to preserve vegetables and fostering an appreciation of the resulting unique tastes and aromas. Many of the techniques involved in preparing *tsukemono* can be adapted for use with local ingredients that are readily available in other parts of the world. A more extensive knowledge of the Japanese preserving methods can serve as an additional inspiration for this revival and inform the evolution of what was until recently considered a 'lost art.'

Vegetables and *Tsukemono*— Made for Each Other

野菜と漬物

Every food culture in all parts of the world, regardless of its ethnic background, turns to the local plant kingdom for many of the ingredients that make up the daily diet. This is where we find fruits, vegetables, tubers, rhizomes, berries, nuts, seeds, legumes, and cereals, as well as large and small herbs. Vegetables, as such, are not a well-defined botanical category and we tend to classify raw ingredients from plants according to how they are used as food rather than on their genetic makeup and morphology. This is why we often refer to mushrooms, large seaweeds, and some fruits as vegetables—think of champignons, kelp, and cucumbers. Similarly, rhubarb is a true vegetable that we think of as a fruit.

Of the hundreds of thousands of different species of plants found on Earth, there are about two hundred that are regarded as vegetables. It is thought that humans started to cultivate plants at least 10,000 years ago and that the first domesticated vegetable was a type of marrow that was grown in the Middle East. The vegetables that we eat today have evolved and been improved by selective breeding over thousands of years to such an extent that they bear little resemblance to their ancestors that grew in the wild. As a result, those that are inedible or poisonous have largely been eliminated from our diet and the individual vegetables themselves are both larger and more nutritious. Unfortunately, in the past few decades the advent of agribusiness-style market gardening on an industrial scale has promoted the production of vegetables that grow quickly, contain more water, and are less flavourful.

Only some vegetables can be eaten raw and many are very perishable once they have been harvested. The question then arises as to how to make prepared vegetables more appealing and this is where storage, preservation, and the culinary arts come into the picture. But before turning our attention to how *tsukemono* can help us to overcome these challenges, let us digress briefly to consider why solving them is of increasing importance.

Moving toward a More Plant-Based Diet

Plant-based food plays a very prominent role in the cuisines of many parts of the world. But in others, including our own, where there is fairly ready access to meat it is either not particularly deeply embedded in the food culture or undervalued. It is to readers where the latter holds true that this section is addressed.

According to nutritionists, we should consume about 600 grams of vegetables and fruit every day, a quantity that is considerably in excess of what many of us normally eat. Their advice stresses the importance of creating vegetable dishes that are palatable and have an interesting texture in order to meet this goal. Apart from their nutritional value and their contribution to our overall well-being, there are many compelling reasons for paying more attention to our intake of plant-based food or, rather, our failure to eat enough of it. Not least among these are the effects of climate change and the realization that the environmental cost of eating an abundance of meat is too great. According to a report published by the EAT-Lancet Commission in 2019, it will be imperative to move to a more plant-based diet in order to provide sufficient, healthy, and sustainable food for a global population that is expected to reach almost 10 billion by 2050 and do so in a manner that eases the toll it takes on the planet. Consequently, we have no choice but to try to promote more sustainable dietary habits on the part of everyone— we simply have to source a greater proportion of our food from the plant kingdom. But can we adapt to eating that much of it, will it appeal to our taste buds, and will we be willing to try a more varied assortment of vegetables and prepare them in new ways?

There are two fundamental reasons why transforming eating habits that are entrenched in some food cultures poses an immense challenge. One has to do with the biology of plants, and the other is firmly rooted in human evolution.

Plants have, to a large extent, evolved to avoid being eaten using what could be described as a smart chemical system in order to survive. As they are held in place by roots, they are unable to run away to escape an enemy. Not being mobile, they lack muscles and, therefore, have much less stored ATP than animals. Because ATP is a source of free nucleotides that interact synergistically with glutamate to elicit enhanced umami taste, plants have less immediate appeal to hungry animals. In addition, their roots, stem, and foliage, as well as unripe fruits, are often bitter, sour, and even, in some cases, poisonous. Hence animals are not as attracted to those essential parts of the plants and they can continue to grow. The opposite is true for ripe fruits, which are generally looked on favourably. They are usually sweet and some, for example, tomatoes, even have umami taste, which is due to their free glutamate content.

Sweetness in a food is an indication that it is calorie-rich, and umami is a signal of accessible proteins, both factors that are important for survival. So, animals seek out ripe fruit, in the process scattering their seeds and helping the plants to reproduce.

Our distant ancestors were fruit eaters and the craving for sweetness conveyed by the aroma of ripe fruits has stayed with us. For more than 2 million years, however, and increasingly after humans started using fire for cooking about 1.9 million years ago, we have also been carnivores. This diet, which was rich in both calories and proteins, was a prerequisite for the evolution of our big and energy-consuming brain. Meat is an excellent source of the free glutamate and free nucleotides that result in umami, a taste that over time we have come to associate with deliciousness.

The legacies of these evolutionary forces have been the twin human cravings for sweetness and umami. This had led to our preferring fruit and meat, when it is available, over a whole range of foodstuffs that we reject as being insufficiently tasty. But we are now faced with the prospect of having to modify our habits and rising to the challenge of increasing our reliance on plants, both to mitigate climate change and to provide food security for a larger share of the world's population. This brings us back to *tsukemono*, which can help us to gain another perspective on vegetables, one small, delicious, crisp bite at a time.

Making Vegetables More Palatable

Mouthfeel is that part of the taste experience which is also referred to as texture. It is actually due to that aspect of the physical structure of the food which can be perceived by the sense of feel in the mouth. Mouthfeel strongly influences our impression of *tsukemono,* and we use expressions such as crisp, crunchy, elastic, plastic, soft, hard, and pliable to describe it.

Vegetables and fruits are an essential source of the food needed to meet our requirements for calories, proteins, minerals, vitamins, macro- and micro-nutrients, dietary fibre, and antioxidants. Nevertheless, there are many people who avoid eating them in the quantities that are sufficient to meet the recommended daily intake of these nutrients. Based on our experiments, we have concluded that this is largely because many people simply dislike the taste of the vegetables that appear on their plates. Perhaps this is because in some Western cuisines they are often cooked until they are soft and tasteless. With more careful preparation they could be more palatable, retain their vitamin content, give off a pleasant aroma, and remain crisp with an interesting mouthfeel. Is there anything worse than mushy green beans and limp carrots or cauliflower florets that have turned into a soggy mass?

Surely there must be some more imaginative ways to prepare vegetables. The simplest and most obvious one is to eat newly harvested vegetables completely raw, an approach that is championed by the raw food movement. This idea is not without drawbacks as many vegetables are tough, hard, and often bitter, the plant defence mechanisms referred to earlier. It can also be difficult to derive full benefit from their nutrients in uncooked vegetables because they are so tightly bound to the plant tissue that it is hard for the human digestive system to extract them. The problem can be partially solved by chopping the ingredients so finely that they are almost like a purée, grating them, or cutting them into julienne strips. As a result, they take on a completely different mouthfeel that, in turn, contributes to a very different sensory impression. Nevertheless, when eaten raw some vegetables will always taste bitter and often lack the sweetness and umami we are primed to crave.

Raw—Completely Raw

The raw food movement is a lifestyle choice that is founded on the premise that one should follow a vegetarian or vegan diet in which food must not be heated to a temperature above 40–42 °Celsius. The mainstays of the diet are vegetables, fruits, nuts, and seeds. The movement seems to be based on the concept that raw food is more authentic and natural for humans and, consequently, healthier. It is also argued that not heating ingredients above that temperature threshold will preserve their enzymes. This is quite correct, but the potential nutritional benefit of this approach may not be what it is claimed to be. Enzymes and the other proteins contained in the food are broken down in the course of their passage from the mouth and on into, and through, the digestive tract. Some nutritionists have, in fact, warned that the necessarily, rather limited selection of foods that can be eaten raw is so lacking in a number of vital nutrients, vitamins, and minerals that followers of this diet put themselves at risk of falling ill. But there is no doubt that raw vegetables are a good supplement for cooked food, especially as they contribute a pleasing texture that is crisp, crunchy, and juicy.

What is the solution? One answer is to turn to the Japanese art of pickling, *tsukemono*, for inspiration. Prepared this way, vegetables exhibit many desirable qualities: they stimulate the appetite, are palatable and flavourful, have an interesting mouthfeel and enticing aroma, are easily digestible, and can be kept for relatively long periods of time. As a bonus, their nutritional value and healthful potential are often enhanced. The vegetables are transformed so that it is a joy and a pleasure to eat them.

This culinary art is rooted in age-old Japanese preservation techniques that work their magic on vegetables and fruits. The basic methods involve marinating in salt, acidic liquids, sugar, alcohol, and/or immersion and possibly fermentation in media such as soy sauce, *miso*, rice bran, sake lees, and a special medium, called *koji*, which contains a microscopic fungus. Although, in principle, these techniques resemble those used in Western cuisines to pickle, marinate, and ferment using lactic acid bacteria, there are a number of significant differences that will be described in a later chapter.

The Many Varieties of *Tsukemono*

漬物とは何ですか

The origins of *tsukemono* go back more than a thousand years to a time when fresh vegetables were not available in the winter. Because these pickles can be just as tasty and crisp as newly harvested vegetables, they conjure up images of summer throughout the year. They also have a characteristic mouthfeel that is often very crunchy. It is said that chewing on *tsukemono* is the second noisiest dining experience in Japan, with the slurping sound made when eating *soba* noodles taking the prize.

The word *tsukemono* covers a long and very varied list of products that are prepared according to many different methods. Some of these require little effort and result in simple foods that will keep for only a few days. An example is *asa-zuke*, or 'quick pickles,' that are lightly marinated in salt and can be made in under half an hour. Other methods are more elaborate and involve steeping the ingredients for days, weeks, or even a whole year. These pickles, such as *miso-zuke*, which are vegetables preserved in *miso* paste, can sometimes be kept for long periods of time.

Nani wanakutomo ko no mono is a long-winded old Japanese proverb expressing the idea that even if you have nothing else, you will probably get by so long as you have *tsukemono*. These pickles could make a bowl with a little rice into a good meal.

Tsukemono can be made from a vast array of vegetables and vary from one region of Japan to another, resulting in numerous local specialties. The differences in approach even extend to the level of individual families. Jealously guarded recipes are often handed down through the generations and the correct way of preparing the pickles is the subject of many spirited discussions. *Tsukemono* are so well integrated into Japanese food culture that regional festivals devoted to them are held every year.

◩ Cucumbers (*kyuri*) marinated in salt are sold as a snack at a Japanese festival.

It is worth noting that, in the Japanese kitchen, *tsukemono* are not just pickled foods that are served arbitrarily as condiments. They are an essential element of Japanese cuisine. On the surface, they have a Zen-like simplicity, but both their tastes and textures are very sophisticated. They are as much part of an ordinary meal as of a formal, classical *kaiseki* dinner. Commercially prepared *tsukemono* are sold in supermarkets in simple plastic bags, but they can also be bought as much sought after, regional and seasonal specialties, elaborately packaged for presentation as highly prized gifts.

Tsukemono play a special role in relationship to cooked rice. In Japan rice is usually prepared without salt and has a rather bland taste. Traditionally, a few pickles were served with a cup of green tea and a bowl of cooked rice at the conclusion of a meal or, possibly, as a snack at afternoon tea time. Things have changed and a small serving of *tsukemono* may make an appearance before, during, or after a Japanese meal. In addition, they are often combined with other types of food. *Oshinko* (pickled *daikon*), for example, are used as a filling in *maki-zushi*.

A Little Bit of *Tsukemono* History

It is probable that the practice of making products similar to *tsukemono* as a solution to the problem of storing and keeping vegetables and fruits for later use arose independently of each other in those countries in Asia where rice is an integral part of the daily diet. These preserves were an easy way to add fresh, delicious tastes and a little colour to a simple bowl of cooked rice, enhancing both its nutritional value and its visual appeal.

Some researchers trace the origins of *tsukemono* back several thousand years to China, but the practice of preserving fresh ingredients with soy sauce and *miso* was known over time in many other places in Asia. The Todaji temple in the old Japanese imperial city of Nara houses a manuscript, written in the fourth century CE, that mentions a fermentation medium consisting of the mould *koji*, salty water, and soy beans and describes techniques for marinating using sake and rice vinegar. The use of a special fermentation medium made with rice bran (*nuka*) is considered to be a purely Japanese innovation. There are reasons to believe that the consumption of this type of pickle, *nuka-zuke*, was instrumental in reducing the incidence of beriberi in Japan. This disease is attributed

***Tsukemono* and *tsukudani*:** Like *tsukemono*, *tsukudani* are also condiments that are often used as topping on a bowl of cooked rice. They were first made during the Edo period (1603–1887) on the island of Tsukuda as a way of preserving vegetables, seaweed, fish, and shellfish. But the method involved is the opposite of that for preparing *tsukemono* as it involves heating. The ingredients are simmered at length in soy sauce, *mirin*, and sugar until they have dried out or are a little sticky. *Konbu-tsukudani* and *nori-tsukudani* are two examples made with seaweed.

Popular tradition in Japan has it that the custom of serving *tsukemono* with meals originated in the tenth century CE. At that time, when noble Japanese families gathered together, they held competitions to see who could correctly identify different types of incense by their smell. In order to sharpen the sense of smell, they ate *tsukemono*, which they referred to as *konomono*, linking the name of the foods to the fragrant smell of incense.

to a lack of vitamin B$_1$ in an unbalanced diet that consists mainly of polished white rice.

It is thought that *tsukemono* had become well integrated into Japanese food culture by the seventh century CE. During the Edo period (1603–1887), stalls and shops that both prepared and sold the pickles made their appearance. In 1836, a wholesale dealer in Edo, as Tokyo was known then, supposedly bragged about offering an assortment of sixty-four different kinds of *tsukemono*.

Most likely it was a sake brewer who promoted the use of the lees (*sake-kasu*) left over from his sake production to conserve and marinate vegetables to make *kasu-zuke*. An interesting parallel development took place in 1902 in Britain when the Bass Breweries launched a product called Marmite, which is produced using the yeast dregs from beer production. Like *sake-kasu*, Marmite contains a large quantity of vitamin B and hydrolyzed protein, which both elicit umami taste.

Ten Ways to Prepare *Tsukemono*

The many different ways of preparing *tsukemono* are a reflection of the need to conserve these fresh ingredients for extended periods of time, up to several years. The advent of modern refrigerators has completely altered the landscape and there are now many types of *tsukemono* that have a much shorter shelf life than earlier versions. In this respect, the tradition of making these pickled foods is similar to that relating to sushi. Sushi (*nare-zushi*) was originally made by fermenting fresh raw fish in layers of cooked and salted rice over a period of several months.

Tsukemono can be made from virtually any kind of vegetable, as well as fruit, seaweeds, fish, shellfish, and even flowers. The different varieties and the techniques related to preparing them are both denoted by the same word, which always ends with *–zuke*. For example, *miso-zuke* refers to both a method and its end product.

It is said that there are about four thousand different kinds of *tsukemono* in Japan and more than a hundred different ways of preparing them. An amusing comparison can be made with the claim that four hundred different kinds of cheeses are produced in France. This overabundance of choice can be overwhelming, especially when compared to the number of pickled products that are common in Western cuisines. Our knowledge of them generally

Mottainai—avoid waste! As in other traditional food cultures, Japanese cooking has been shaped by the omnipresence of food scarcity, combined with the desire and necessity not to let anything go to waste. This is inherent in the idea of *mottainai*, a Bhuddist concept that conveys feelings of regret over wastefulness or the misuse of foodstuffs. Preparing *tsukemono* is a tangible manifestation of this attitude by making it possible to keep vegetables for consumption outside of the growing season. Moreover, basically everything can go into the rice bran pickling bed (*nuka-doko*)— the roots, the stems, and the green leaves of the vegetables. And, naturally, the fermentation and marinating media, be they *nuka-doko*, miso, or soy sauce, are used over and over again.

covers a much more limited range, which includes various types of pickled cucumbers, pickled beets and red cabbage, sauerkraut, pickled herring, pastrami, and capers, where the dominant taste impressions are sweet and sour, with the addition of some spices.

◘ Different types of *tsukemono* for sale at a market in Japan.

Tsukemono are often classified into two different categories that are distinguished by the duration of their pickling time. Those that can be prepared quickly, for example, *asa-zuke*, are marinated in salt or in weak brine for a few hours or days. They usually need to be kept refrigerated and are eaten within a few days. The others, called *furu-zuke*, require long, elaborate preparation, conservation, and ageing for periods that range from a few weeks to a year or more, and will keep for months and up to several years. As long as they are stored in a dark place and in the original pickling crocks to keep them from drying out, they do not need to be cooled. The differences in preparation and ageing times are reflected in distinct variations in flavour and fragrance. A common trait, however, is that most *tsukemono* made from vegetables are crisp and so crunchy that every bite releases a veritable explosion of aroma and taste, in particular umami. Some Japanese chefs describe this experience as a taste with long duration, much as the French use *longueur* to describe the finish and aftertaste of wine.

The various techniques for preparing *tsukemono* can be divided into ten basic types that are sometimes used in combination. These types correspond to three different procedures—salting and marinating in liquid, preserving in a paste, where there may also be a simultaneous fermentation process, and a genuine fermentation process mediated by enzymes and microorganisms such as fungus, yeast, and bacteria. With all three methods, the addition of salt, either directly or indirectly from other ingredients, is the most essential element. If a paste is involved, it is normally wiped off, either completely or partially, before the preserve is eaten. Both the liquid marinade and the paste can be flavoured with spices and additives, such as Japanese chili (*togorashi*), Japanese pepper (*sansho*), mustard (*karashi*), ginger, *yuzu*, *shiso*, and seaweeds (*konbu, wakame, nori, ao-nori*). The preparation technique for a particular vegetable or fruit is chosen in order to enhance its distinctive character, texture, and taste.

The simplest way to make *tsukemono* is basically just to marinate the ingredients in brine containing 5–25 percent salt. Ordinary table salt (NaCl), which has a dependably uniform composition, is not used exclusively. In Japan it is traditional to choose instead from a variety of sea salts, a topic to which we will return later on. These may contain additional salt compounds, for example, potassium, magnesium, and calcium salts, as well as substances, such as seaweed ash, that are actually desirable impurities. It is, therefore, crucial to select the right type of salt as these differences influence the outcome of the pickling process, with respect to both the taste and the texture of this type of *tsukemono*.

The recipes for the different types of *tsukemono* are generally simple. In the case of the ones that require long-term ageing, it is not so much a question of time as it is of patience.

◘ *Daikon* before and after having been marinated in brine for several weeks.

It is, by far, easiest to work with liquid marinades consisting, for example, of brine, soy sauce, sake, or vinegar. Using fermentation media containing pastes and fungus cultures is more complicated, but the reward is a considerably more nuanced taste. Here it is truly possible to talk of *konomono*, 'aromatic things with a very delicious aroma.'

In some instances, several preservation media, with different effects, are used in combination. For example, a marinade might consist of soy sauce, sake, and sugar or might be a mixture of *miso* and sake lees.

◘ A selection of ingredients used to make *tsukemono*: wheat bran, rice bran (*nuka*), sake lees (*sake-kasu*), *miso*, and *koji*.

In the case of vegetables, dehydrating them before they are preserved results in a much crisper and crunchier mouthfeel. Those that are to be immersed in a thick paste, generally for several weeks or months, absolutely must be desiccated first, otherwise the preservation medium becomes too moist. Dehydration can be carried out using salt, according to old-fashioned methods out of doors, or with the help of a dehydrator.

Sunomono, which are simple salads often tossed with a slightly sweet vinegar dressing, are closely related to *konomono*. A common one consists of cucumbers mixed with a little seaweed, usually *wakame*.

Sunomono can be translated as 'things that can be prepared in vinegar (*su*).' They are often in the form of small salads made with cucumbers or other greens in a tart marinade. We would probably think of them as a type of side dish, but in Japan one would call them *hashi-yasume*, which literally means 'vacation for the chopsticks,' implying that the chopsticks should have a rest while one eats a very small portion of *tsukemono*. This is analogous to the mini servings of palate cleansers that are interspersed between the courses of an *haute cuisine* meal, especially meat dishes and others that have a high fat content.

1.   Marinating in salt and brine (*shio-zuke*)

Marinating in salt is the simplest way to make *tsukemono*, using virtually any type of vegetable and certain fruits. Those which are just sprinkled with salt, known as *asa-zuke*, will keep for a few days in the refrigerator. The salt helps to draw out the bitter substances that are common in many green vegetables and acts as a temporary preservative. A strong salt solution, that is, a brine, is sometimes the starting point for making fermented *tsukemono*.

Shio-zuke prepared from cucumbers and carrots are the most common, but there are practically unlimited possibilities for using cabbage, broccoli, cauliflower, aubergines, bell peppers, and onions. Placing a few pieces of *konbu* in the pickling crock adds sweetness and brings out umami. *Shio-zuke* made with whole cucumbers (*kyori asa-zuke*) are very popular in Japan, where they are cut up and served as a refreshing side dish with a meal, as well as eaten whole as an outdoor snack at festivals.

Hakusai-zuke are *shio-zuke* made from Chinese cabbages (also called Napa Valley cabbages), often mixed with carrots and cucumbers and seasoned with *konbu*, *togorashi*-chili and grated *yuzu* peel. *Hakusai-zuke* are eaten as a salad and as a complement to fish dishes.

◘ *Nasu shio-zuke* made with aubergines.

Umeboshi are probably the most traditional type of Japanese *shio-zuke*. *Ume* are small stone fruits, resembling apricots, that are dried and preserved with salt and red *shiso* in a plum wine marinade. These fruits contain a great deal of citric acid, which impedes the growth of undesirable microorganisms and aids in digestion. *Umeboshi* have a sharp sour-salty taste that combines with aromatics from the red *shiso*. Their texture is soft and meaty.

Ume (*Prunus mume*) are small stone fruits that are related to both plums and apricots, but more closely to the latter. Confusingly, the fruit is often referred to as a Japanese apricot, but the wine made with it, *umeshu*, as plum wine.

◘ Preparation of *hakusai-zuke* from Chinese cabbage.

◘ Dehydration and marinating of *ume* for the production of *umeboshi* with red *shiso*. *Umeboshi* are a typical *shio-zuke* with a sour and aromatic taste.

In season, unripe *ume* can be made into a crisp, crunchy *shio-zuke*. The small green fruits are placed in brine and are often eaten as a snack.

▪ *Ume shio-zuke*: brined *ume*.

2. Marinating in a vinegar solution (*su-zuke*)

As Japanese rice vinegar (*su*) is normally not very acidic, *tsukemono* preserved this way, with the exception of those from leeks and Chinese onions (*rakkyo*), will keep under refrigeration for only a few days.

Beni-shoga is made from fresh shoots or slices of the root of ginger (*shoga*), sprinkling it with salt, and then marinating it in *ume* vinegar (*umezu*) and, possibly, some marinade left over from making *umeboshi*. A naturally occurring pigment in the ginger (anthocyanin), causes it to take on a red colour when it comes in contact with the vinegar. In contrast to *gari*, the other kind of pickled ginger root that is used as a condiment with sushi and sashimi, *beni-shoga* is not sweet. Because of its beautiful red colour, it is used as decoration on other dishes. It is quite sour, with a crunchy, crisp texture.

3. Marinating in a sugar solution (*sato-zuke*)

Sato-zuke are distinctly different from other types of *tsukemono*. No salt is used, and the ingredients are simmered for several days in a sugar solution. This crystallizes them, much in the same way as is done with candied fruit in the Western tradition. Melons, lotus root, and ginger root, as well as *ume* and *yuzu* peels are commonly prepared this way.

▣ *Yuzu sato-zuke*: sweet, crystallized *yuzu* peel.

4. Marinating in rice vinegar and sugar solution (*amasu-zuke*)

Gari, sweet and sour pickled ginger root, is probably the type of *tsukemono* that is the most widely known outside of Japan, because it is almost always served with sushi and sashimi. Finely sliced pieces of young ginger root are first salted and then placed in a solution of rice vinegar and sugar (or *mirin*). The ginger turns a pinkish colour in the presence of the vinegar. Commercially produced *gari* is often coloured with red *shiso* or food colouring.

The fresh, slightly astringent taste of *gari* makes it very suitable for cleansing the palate between bites of the various pieces of sushi, where there are subtle differences in the taste of an assortment of fish and shellfish. It has a clean, sweet, and grass-like taste, is slightly peppery, and has a crunchy texture.

◘ *Gari*: Sweet and sour pickled young ginger root, best known as a condiment for sushi and sashimi; here presented with a little pickled cherry blossom.

Senmai-zuke are a Kyoto specialty. Paper-thin slices of young turnips are placed in a barrel, together with *konbu*, in a marinade made with sweet rice wine (*mirin*). Sometimes this type of *tsukemono* is seasoned with Japanese chili (*togorashi*). The seaweed, which is also eaten, causes the liquid that is drawn out of the turnips to thicken slightly and take on a wonderful sweet taste with a great deal of umami. *Senmai-zuke* have a crunchy mouthfeel and are slightly sticky and slimy due to the carbohydrates that have been drawn out from the seaweed. The slime should not be washed off as it is a component of the overall taste impression. *Senmai-zuke* taste a little sweet, a little salty, a little sour, and a bit like cabbage. Originally *senmai-zuke* were also fermented.

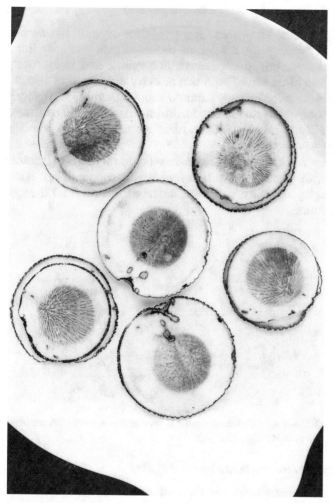

■ *Senmai-zuke*, here made from black radishes, which are sliced thinly and preserved in salt and *mirin*; circular pieces of *konbu* are placed in between the slices.

5. Marinating in soy sauce (*shoyu-zuke*)

The best known and most common type of *shoyu-zuke* is *fukujin-zuke*, a relish or chutney made primarily from *daikon*, brined cucumbers, aubergines, and lotus roots, and often served with rice or curry dishes. Other vegetables—sword beans, bamboo shoots, *shiitake*, and ginger root—are sometimes added to the basic four ingredients. *Fukujin-zuke* tastes like a sweet chutney and has a crunchy and crisp texture.

Another type of *shoyu-zuke* (*kyuri shoyu-zuke*) is made with cucumbers and seasoned with pepper. They taste spicy and salty, with a great deal of umami from the soy sauce.

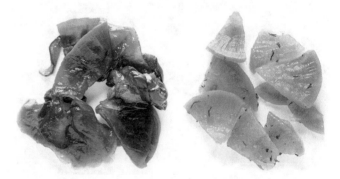

◘ *Shoyu zuke*: *daikon* marinated in soy sauce with different seasonings and possibly with other vegetables.

6. Preserving in rice bran (*nuka-zuke*)

Rice bran (*nuka*) is the part of the rice kernel that is removed when rice is polished. The bran has a high content of proteins, fats, and carbohydrates, which makes it a good medium for fermenting and making *tsukemono*. *Nuka-zuke* are often prepared in large portions as they are pickled for a long period of time, typically for more than three months.

◘ Large assortment of *nuka-zuke* at a Japanese market.

The taste of *nuka-zuke* depends to a great extent on the specific rice bran bed (*nuka-doko*) that is used, and people have their individual preferences. Their taste is somewhat salty, astringent, and a little sharp, often with slightly earthy overtones, as if they have been kept too long.

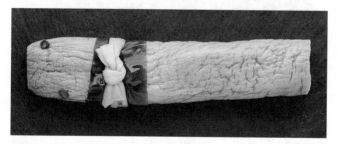

◘ *Takuan-zuke*: *tsukemono* made from *daikon* placed in a rice bran fermentation medium (*nuka-doko*).

Takuan. It is said that *takuan-zuke* were invented by a Buddhist monk by the name of Takuan Soho in the 1700s. For many Japanese, *takuan-zuke* are the very essence of traditional, old-fashioned *tsukemono*. As a result of the long fermentation period, they have both a strong taste and smell, which many find off-putting. They can be difficult to prepare because it is challenging to find the right balance of humidity and temperature in the fermentation bed. If the *daikon* dry out too much, they become hard.

Takuan-zuke, made from *daikon*, are the best known type of *nuka-zuke*. They are served as a condiment with many different dishes that include cooked rice and are often used in vegetarian *maki-zushi* (*oshinko-maki*). The long radishes are sliced before they are served. *Takuan-zuke* are very crunchy and crisp, but also a little juicy. The taste is somewhat sweet, sour, and a bit sharp and prickly. The mouthfeel is hard and crunchy.

When the vegetables are removed from the fermentation bed, they will keep for only a short period of time, unless they are vacuum packed, or preservatives are added to them. Normally, *nuka-zuke* are washed just before they are served.

◻ *Maki-zushi* with *takuan-zuke* (*oshinko-maki*).

A specialty called *iburi-gakko* from the Akita Prefecture in Japan is prepared from *daikon* that are first smoked, often with cherry, apple, or chestnut wood, and then placed in *nuka-doko*. This technique was probably invented on account of Akita's cold climate, where it is difficult to dry *daikon* outdoors during the autumn. Using smoke to dehydrate the vegetables both reduced the water content and helped to preserve them, in the process adding an interesting taste element.

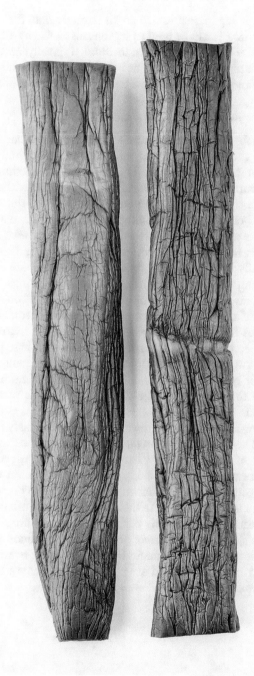

◘ *Iburi-gakko* are a local *tsukemono* specialty made from smoked *daikon* in the Akita Prefecture of Japan. The indentation in the middle of the *daikon* is caused by the string that was used to hang up the vegetable while it was being smoked.

7. **Marinating in strong mustard (*karashi-zuke*)**

Karashi is Japanese mustard and is stronger than the ordinary variety. It can be used to season *tsukemono* by adding it to a rice bran fermentation medium. Typically, *karashi-zuke* are made with aubergines or zucchini. Their taste is intense and burning and their texture is elastic and pliable.

8. **Immersion in sake lees (*kasu-zuke*)**

Kasu-zuke are made according to a technique that is at least a thousand years old and evolved in conjunction with the brewing of sake. The fermentation medium is made from sake lees (*sake-kasu*), possibly mixed with sake or *mirin* to obtain the right consistency. Sugar, salt, ginger, and citrus juice, from either lemons or *yuzu*, may be added as well. Sometimes *miso* is stirred into the *sake-kasu* and this has a tendency to speed up the ageing process. Normally it takes months or even years to prepare *kasu-zuke* from vegetables, but they are very long-lasting.

If one is simply looking for a milder taste, one can make delicious *tsukemono* by placing the ingredients in sake lees for a few hours. Usually most of the marinade is scraped off before the *kasu-zuke* are eaten. These pickles are sweeter and less salty and sour than the majority of other types of *tsukemono*, but the taste becomes more intense the longer the vegetables are left in the pickling medium.

Nara-zuke are among the most famous and tasty *kasu-zuke*. They are usually prepared in large batches with a variety of vegetables that have a nice texture, especially cucumbers, marrows, and Japanese pickling melon (*uri*). The vegetables take on a darker hue according to how long they are left in the lees. As their taste is intense after they have been cured for a long period of time, they are served in thin slices. *Uri nara-zuke* are the most common type. They have a sweet, aromatic taste with strong overtones of sake and alcohol, and are among the crunchiest of all *tsukemono*.

▣ *Nara-zuke* made from Japanese pickling melon (*uri*) immersed in *sake-kasu*.

Wasabi-zuke is a specialty of the Shizuoka Prefecture in Japan where *wasabi* grows in abundance. It is, made from the stalks and blossoms from *wasabi* plants that are marinated in *sake-kasu*.

Sake-kasu is also used to prepare fish. After being immersed in the medium for a few hours the fish is eaten raw, steamed, or fried lightly on a pan.

9. Immersion in *miso* (*miso-zuke*)

Miso-zuke are prepared according to a technique that is just as old as the one for *kasu-zuke* and is also the one that can stretch over the longest pickling period. Typically, *miso-zuke* are made with Japanese pickling melon (*uri*), garlic, burdock roots, marrows, or tofu. How they turn out is very dependent on the type of *miso* used. The taste is stronger when red or dark *miso* is used for the *miso-doko* and depends on whether the *miso* is made from soybeans only or from a variety of grains, such as barley, brown rice, and buckwheat. Red *miso* (*aka-miso*) is preferred as it results in a somewhat stronger taste than that from the sweeter white *miso* (*shiro-miso*).

The 'holy trinity' of classical *tsukemono* in Japan consists of *umeboshi, takuan-zuke,* and *hakusai-zuke.*

It is also entirely possible to make tasty *miso-zuke* from vegetables that have been marinated in a single type of *miso* for only a few days by placing them with other vegetables and spices that have already been immersed in a pickling medium (*miso-doko*) consisting of different types of *miso*, often mixed with sake, for a longer period of time.

The different types of *miso-zuke* are often used in sauces, soups, and hearty warm dishes because they have such a strong taste. They are also used to accompany fish and shellfish.

10. Fermentation with *koji* (*koji-zuke*)

The preparation of *koji-zuke* calls for a particular fermentation culture called *koji* that, among other ingredients, contains microscopic fungi of the *Aspergillus oryzae* variety. *Koji* is also used for the production of soy sauce, *miso*, and sake.

Koji-zuke usually take on a sweet taste because the fungi secrete the enzyme amylase, which converts the starch in vegetables into sugar. Other enzymes break down proteins and form tasty free amino acids resulting in umami taste, while lipases break down the fats. The overall effect of *koji* on the food is not only to impart a more delicious, complex taste, but also to make it easier to digest.

Tokyo is the home of a classical type of *koji-zuke*, *bettara-zuke*, made from *daikon* immersed in a marinade of sugar, salt, and sake together with *koji*. When the *koji* fungus grows, it forms a slimy network that is also eaten. Unlike *takuan-zuke*, which are prepared using dehydrated *daikon* and are firm and crisp, *bettara-zuke* have a sticky, wet mouthfeel.

Shiba-zuke Kyuri-zuke Daikon nara-zuke Shiba-zuke

Takuan-zuke Fukujin-zuke Matcha takuan-zuke Senmai-zuke

Kabocha-zuke Iburi-gakko Gari Nasu nara-zuke

Nasu shio-zuke Shiba-zuke Takana-zuke Uri nara-zuke

Kyuri-zuke Kabu shio-zuke Shiba-zuke Shiba-zuke

Takana-zuke Shiba-zuke tsukedani Shiba-zuke Takana-zuke

Salt, Taste, Mouthfeel, and Colour

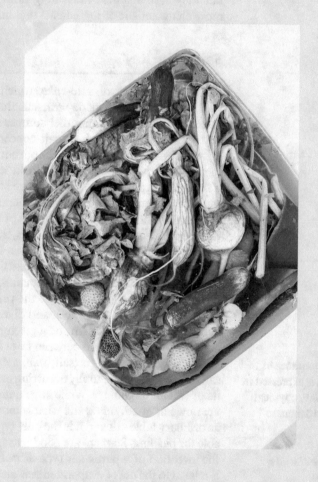

There are so many different ways of preparing *tsukemono* utilizing a wide range of vegetables and fruits that it is not possible to make many general comments about their tastes and colours. Nevertheless, they have something in common, which is mostly due to the techniques employed to preserve them. These normally require the use of salt and often involve dehydration as well. Because making *tsukemono* never involves the use of heat, the mouthfeel of the pickled vegetables is as crisp as when they were still raw, but they have acquired interesting taste nuances that are more characteristic of ingredients that have been cooked.

Salt Is the Key

With the exception of *sato-zuke*, which are sweet, every type of *tsukemono* is made with salt that is added either directly or indirectly from other sources such as soy sauce and *miso*. Originally, a high salt concentration, often as large as 8–10 percent, was an important factor in ensuring that the pickles would keep for a long time. And as a further precaution, the pickles were stored in the coldest, darkest part of the house. Refrigeration has radically altered the picture. With the addition of preservatives and the use of effective cooling techniques, the salt content has now been reduced to 3–4 percent, even for the *tsukemono* that are intended to have a long shelf life. Some modern varieties are also pasteurized, vacuum-packed in plastic bags, and kept refrigerated. Here the primary role of salt is its effect on taste, mouthfeel, and digestibility. In all circumstances, the salt content of a given product is a question of striking a balance between taste and conservation.

The importance of brining is expressed in this *haiku*, a typical stylized Japanese poem.

pickled in salt
now it will send a
message
the imperial gull
Matsuo Bashō
(1644–1694)

In Japan, the word for salt, *shio*, actually refers to sea salt, as there are no naturally occurring underground deposits of salt in the country. While large quantities of rock salt are now imported, *shio* is still what is most commonly used as ordinary table salt and it is the only type considered suitable for pickling. So there are good reasons to suppose that the invention of a particular type of *tsukemono*, *shio-zuke*, is related to the use of evaporated seawater as a preservative.

Due to the generally very humid climate in Japan it is not feasible to evaporate seawater efficiently by exposing it to sunlight in shallow land-based ponds, as is done in Mediterranean areas. Consequently, the Japanese, like many other peoples, produced salt early on by burning seaweeds, placing the resulting ashes together with seawater in a kettle, and then retrieving the salt by evaporating the water over a fire. This type of salt is called *moshio* and has a strong taste of the sea and of seaweeds.

In the ninth century CE a very laborious technique, called *agehama*, came into use. First, seawater was repeatedly poured over sandy paddies on the beach to evaporate some of the water. Next, the salt-coated sand was placed in a tank and more seawater was added, thereby increasing the salt content. The resulting briny liquid was then heated in special drying ovens to extract the salt. A more efficient and less labour-intensive technique was introduced at the beginning of the 1700s. It employed tidal movements to wash seawater over a series of vertically displaced terraces where both sun and wind played a part in the evaporation process.

In order to ensure sufficient salt supplies, the production of salt was under strict government control from 1905 to 1997. By 1972 severe restrictions were placed on the use of the old-fashioned techniques both to save energy and by the need to make many parts of the coastal shores available to accommodate the rapid growth of industrialization. Mass production of salt was converted to make use of ion exchange equipment, which can isolate sodium chloride (NaCl) from the other salts in the water. Interestingly, the authority in charge of salt production regulations also controlled tobacco sales. In 1997 government restrictions regarding the production of sea salt were lifted and there are now more than 1500 small artisanal salt works in Japan.

There is a whole range of different types of sea salt. Each has a distinctive taste related to the coastal area where it is harvested and to its content of mineral salts other than NaCl. These highly desirable 'impurities,' some of which have a bitter taste, consist primarily of calcium, potassium, and magnesium and can make up as much as 12–13 percent of the overall salt content. These variations determine how the final product will taste. The colour of the salt can be a reflection of some residual bits and pieces of algae and seaweed. When used to prepare *tsukemono*, the size and shape of the individual salt crystals are unimportant as they are dissolved in the pickling liquid and do not contribute to mouthfeel. But the content of calcium and magnesium has an important effect on whether the pickles will have a firm and crisp texture.

The sparing use of salt for preparing *tsukemono* and to season simple, everyday foods such as cooked rice, soy sauce, and vegetables, is closely tied to its place in Shinto, the traditional Japanese religion that reveres all things natural and their spiritual dimension. Salt is regarded as a gift from the gods and its use ensures humans of having good health.

◘ Traditional reclamation of salt on a Japanese coast. Wood block print by Utagawa Hiroshige (1797–1858).

Taste and Mouthfeel

When one thinks of typical sweet and sour pickles, such as bread and butter pickles and pickled beetroots that are common in northern Europe and North America, subtle tastes do not immediately come to mind. The relative balance among the basic tastes of sour, sweet, and salty is very prominent and calls forth a sharp, pungent sensation. Many find this adds freshness and provides a contrast to dishes that might be rich, heavy, and have an abundance of umami. Other taste impressions are often due to the presence of popular spices such as pepper, vanilla, chili, and dill.

Tsukemono can be much more interesting than their Western cousins. When properly made, especially if they are fermented, they exhibit an abundance of subtle taste notes. And even though they may impart sharp, distinctive tastes, seasonings such as *togorashi*-chili, *shiso*, and *yuzu* are associated with more complex aromas than those from pepper, lemon, and dill. The addition of liquids, for example, sweet *mirin*, *shochu* (a distilled liquor), *ume* vinegar, and alcohol from sake lees (*sake-kazu*) can further enhance tastes and help to impart aromatic substances.

The salt used to prepare *tsukemono* draws liquid out of the vegetable cells and destroys them. This releases their enzymes, which assist in the process of degrading many of the bitter substances present in raw vegetables, leaving them with a milder, sweeter taste. At the same time, microorganisms, such as yeast and bacteria, begin their work of breaking down the plant matter. This brings about the formation of a whole new series of taste substances, counteracts the impression of saltiness, and leads to rounder, softer taste nuances. In addition, the salt prevents the growth of unwanted bacteria.

As noted previously, *tsukemono* are extra crisp if they are preserved using sea salt, as its content of calcium and magnesium salts cause the carbohydrates in the vegetables, for example, pectin, to form strong bonds, a process known as cross-binding, resulting in a very firm mouthfeel. This is exactly the same effect used to make jelly and other hydrogels from alginate, a gelation agent extracted from seaweeds. The importance of texture is evident from the fact that the Japanese language has seven different expressions for 'crisp.'

The Colour of *Tsukemono*

The colour of *tsukemono* has significance for its esthetic appearance. In some cases, the original colours of the vegetables and fruits are virtually unchanged, for example, those of *shio-zuke*. In other instances, the colours are altered or lost by contact with salt, acid, marinade, or a fermentation medium and from ageing.

Immersion in *sake-kasu* and *nuka-doko* results in pickles in warm shades of yellow and light brown. *Shoyu-zuke* and *miso-zuke* take on similar or even darker colours, especially if red or dark *miso* is used. The longer the *tsukemono* are left in the preserving medium, the darker the colour, and the original colour of the vegetables often fades away. Brown tones are due to some particular substances produced by low-temperature Maillard reactions, which are caused by processes of relatively long duration that combine proteins and carbohydrates during fermentation.

Red *shiso* (*Perilla frutescens*) is used as a natural red dye, which has anti-bacterial properties and also acts as an effective preservative, in *umeboshi*, *gari*, and *shiba-zuke*. Artificial colour additives are frequently used in the commercial production of these types of *tsukemono*. The reduced salt and acid contents, which are now more common in quickly prepared varieties that are intended to have only a short shelf life, have less effect on the original colour of the ingredients, so it is not necessary to add food colour.

Spices and Other Flavour Enhancers

A variety of spices and other flavour enhancers can be added to both the marinades and the fermentation media used for making *tsukemono*. Commonly used ones include *sansho* and *togorashi* peppers, ginger, *yuzu*, red *shiso*, *yukari* (salt-preserved red *shiso*), and seaweeds such as *konbu*, *wakame*, *nori*, and *ao-nori*.

Fronds of *konbu*, which are often placed in the marinade, affect the process in two ways. Certain polysaccharides are drawn out of the seaweed and help to make the liquid a little more viscous, thereby affecting its mouthfeel. In addition, *konbu* contains an abundance of free amino acids, which contribute umami. One of these is monosodium glutamate (MSG), which dissolves readily in the marinade. This infusion of umami has the important effect of enhancing the taste of the *tsukemono* to the extent that it surpasses that of the vegetables from which they were made.

◨ Assortment of *tsukemono* in a 5 percent brine with a variety of seasonings (hibiscus, saffron, curry, *wasabi*, mushrooms, and neutral).

Monosodium glutamate (MSG) is added to many commercially prepared *tsukemono*. It enhances sweet and salty tastes and is the source of umami. In contrast to what is claimed by many, there is no scientific basis for the belief that monosodium glumatate poses a danger to health. The monosodium glumatate produced in factories is identical to the naturally occurring substance found in large quantities in seaweeds, shellfish, and ripe tomatoes, and that are formed by the fermentation processes involved in ripening cheeses and making soy sauce, *miso*, sake, and fish sauce.

Techniques
and Methods

The techniques used to prepare *tsukemono* are basically the same as the ones used to conserve other foods so that they do not deteriorate and become inedible, or possibly even spoil to the point of being toxic. The various methods employed serve not only to extend the keeping qualities of foods, but also to change their tastes and mouthfeel. An edible ingredient, for instance, a vegetable, is exposed to the risk of being broken down on two fronts. It may be attacked from the outside by microorganisms such as bacteria and fungi. And as soon as it is harvested, the enzymes that are naturally present within its cells threaten to start degrading both its physical structure and its constituent substances, including fats, proteins, and carbohydrates.

To a large extent it is possible to delay the onset of both of these problems by cooling or, even better, freezing the foodstuffs. Unfortunately, neither the enzymes nor the bacteria are rendered completely harmless by freezing and they can quickly become active again when the foods are thawed. In addition, freezing can damage and significantly alter their texture.

The external attacks can best be prevented by killing the microorganisms or removing the substrate on which they can thrive. It is easiest to kill them with heat, for example, by pasteurizing them, and then sealing the food in packages, jars, or cans. Smoking is another way to eliminate, to a certain extent, the bacteria and fungi as they cannot withstand the effect of the toxins in the smoke. Alcohol and acidic liquids weaken the substrate for the microorganisms making it harder for them to survive and, as an added bonus, alcohol acts as a dehydration agent.

With regard to internal attacks on the structure of the food caused by enzymatic activity, it is necessary to denature the enzymes or create conditions in which they cannot function. Heating is an effective tool for denaturing the enzymes.

The most effective technique, however, is dehydration as it works on both problems at the same time. It eliminates the substrate that is vital for the microorganisms and reduces enzymatic activity. Removing water preserves the nutritional value, vitamins, taste, and mouthfeel of the original raw ingredients and is, in many ways, a gentler means of handling them. More about this later.

To prepare some types of *tsukemono* that are fermented it is absolutely crucial to use *koji*, a very special fermentation medium. *Koji*, which originated in China, contains certain microscopic fungi with enzymes that are able to break starch down to sugar, which fuels the activity of yeast cells. This process yields some alcohol and creates a vast number of different taste substances. These are responsible for the umami that is characteristic of products made with *koji*, including soy sauce, sake, and *miso*. In recent years, *koji* has been incorporated into modernistic cuisine where it is used to ferment fish, meat, vegetables, legumes, pulses, grains, and insects.

Koji

Koji is made from a solid mass of either cooked soybeans, rice, barley, or another type of grain, seeded with the spores of a microscopic fungus, usually *Aspergillus oryzae*. It can be a bit confusing that the term *koji* refers both to the spores and to the mass in which they sprout. The fungus sprouts from a mycelium that grows in the solid for about three days at a temperature of 30 °Celsius in a very humid environment. At this point, the *koji* can be dehydrated and stored for later use, for example, in the form of grains of rice seeded with the fungus spores. When it is to be used, it is placed in a saline solution together with the ingredients that are to be fermented. As a result, a yeast culture either starts to grow spontaneously or the process is set in motion by introducing an established yeast culture or lactic acid bacteria, which thrive in these surroundings, into the mixture. The activity of the hydrolytic enzymes and microorganisms breaks down proteins and fats, creating a whole arsenal of taste and aromatic substances, among them free amino acids and a significant amount of glutamate. *Koji* is used in the production of soy sauce (*shoyu*), *miso*, and sake.

In principle it is possible to start a *koji* culture on the basis of spores that are found on plants, for example, rice plants, or in the air surrounding organisms that are undergoing fermentation. It can be difficult to ensure that this will lead to consistent results, however, as the different types of fungi can vary significantly depending on the locale and other factors. For this reason, the Japanese have been developing and producing *koji* in laboratories with a controlled environment since the Meiji period (1868–1912). There are now seven such establishments, which manufacture two important strains—one that is used for *shochu* (a distilled alcoholic beverage) and another that is suitable for making sake, *amazake* (a sweet drink with low or no alcohol content), and *miso*. The fungus spores needed to start a *koji* culture are available commercially either as a freeze-dried product, *koji-kin*, which is a mixture of spores and rice flour, or as a salty emulsion of spores and rice or a filtered version thereof, *shio-koji* and *iki-tai shio-koji*, respectively.

Koji made with malted rice, as *shio-koji*, and in liquid form as *ikitai shio-koji*, respectively.

The different conservation techniques affect the structure of the raw ingredients and, with it, their mouthfeel, in a variety of ways. The important challenge is to employ these methods so that the resulting *tsukemono* are so crisp that one can hear them being crunched loudly between the teeth.

The Physical Structure of Vegetables

In order to understand what is required to prepare *tsukemono* with a particular mouthfeel it is necessary to take into account the composition of vegetables at the cellular level. The combination of the inner structure of the cells and of the cell walls and the way in which they are bound to each other has a significant effect on their texture. In addition, the influence of moisture content (water or juice), salt, acid, and temperature on cellular construction completely determines how the pickles will feel in the mouth and what it will be like to chew on them.

Plant Cells

In contrast to animal cells, plant cells are protected by a cell wall composed of cellulose. This creates the stiffness and support that enables a collection of cells to form a structure such as a leaf or a stem that can hold itself upright. Cellulose is a carbohydrate, which is not water soluble and which humans cannot digest because our bodies do not have the enzymes that can break it down. This is why cellulose is classified as an insoluble dietary fibre.

Other dietary fibres, which are soluble, as well as water and minerals are found between the cells. These types of fibres are exceptionally good at absorbing water to form a semi-solid that we call a gel. In a sense, this is the glue that binds the cells together. Some of these fibres are made up of the insoluble carbohydrate hemicellulose. Others are composed of pectin, a soluble carbohydrate that very effectively binds water. This is a well-known property on which we depend, for example, to make jelly from fruit juice. Pectin typically makes up 15–20 percent of the fibre content in vegetables. The fibres hold the cells together by binding the cell walls to each other in all directions.

The essential energy depot for vegetable cells is found inside them in the form of starch, which is a carbohydrate. In general, vegetables contain only a little fat. When they are heated and cooked, the starch gelatinizes, the cell walls become less stiff, and the plant tissue develops a texture that is soft and chewable. The cellulose is, however, not broken down completely and, therefore, cooked vegetables still retain a lot of their insoluble dietary fibre.

Nevertheless, the most important factor affecting the crispness of a vegetable is that the cells each contain an organelle called a vacuole. The vacuole can occupy up to 80 percent of the volume of the cell. Even though the name 'vacuole' would lead one to believe that it is an empty shell, this is far from the case. The cell of a juicy fresh vegetable can consist of as much as 90 percent water, most of which is in the vacuole. While the vacuole is the water depot of the cell, it also carries out other tasks such as storing nutrients and breaking down waste products.

Turgor and Crispness

The rigidity of plant cells, known as turgor, is controlled by what is called the osmotic pressure, which is formed across the cell membrane that lies under the cell wall and across the inner membrane that surrounds the vacuole. Osmotic pressure is a fundamental physical-chemical effect that can arise over a membrane that is permeable only to small molecules, such as those of water, and impermeable to large molecules, such as those of salt, sugar, and proteins. The side of the membrane on which most of the small molecules are located has a tendency to draw water through the membrane, giving rise to a difference in pressure. This, in turn, is responsible for turgor. If a living vegetable does not have a sufficient supply of water or if we try to draw water out of a harvested one by drying or marinating it, the osmotic pressure is reduced, and the vegetable becomes limp. It is possible to reverse this process to a certain degree by adding water. If, in the meantime, the cell wall has been damaged by being heated or frozen, the cell can no longer maintain osmotic pressure and the liquid seeps out and cannot be drawn in again.

As long as the vacuoles are brimful a vegetable or fruit will remain turgid. When its cells are ruptured by chewing action the liquid bursts out of the vacuoles, leading to the mouthfeel we call juiciness. Crispness can be defined as the resistance of the vegetable to the actions of the teeth, which are able, if the bite is sufficiently strong, to fracture the cells causing the juice in them to spurt or seep out. It is the mouthfeel of resistance and of fracturing, together with the release of liquid, that determines whether or not we consider a cucumber or an apple to be crisp and juicy.

Making Vegetables Crisp Again

Fresh vegetables that have become a little soft and limp because of water loss can be made crisp again by immersing them in ice-cold water. The fastest way to do this with vegetables such as carrots or radishes is to peel and julienne them or cut them into thin slices before they are immersed in the water. After a short time, the cells of the vegetables will have drawn in some of the water and the vegetable pieces will be firm and turgid.

◻ Julienned raw carrots, green asparagus, and *daikon* that have been immersed in ice-cold water.

If the cells have a low water content, for example, during a period of drought, turgor will be reduced and the vegetables will feel limp, a little tough, and less juicy. Crispness also depends on the composition of the cell walls, which are largely made up of the molecules of an insoluble fibre, cellulose, that are held together by hydrogen bonds.

When a vegetable or a fruit ripens, its pectin content might be broken down by the action of an enzyme called pectinase. This loosens the bonds between the cells so that when one chews on the vegetable or fruit it will feel mealy and dry, even though its water content may not have changed at all.

As a plant grows, the cellulose fibres also become cross-linked by lignin, an insoluble polymer that is deposited on the walls of some cells, in particular those that are vital for water transport in the plants. Lignin reinforces the cell walls, providing the structural strength that is characteristic of trees. Vegetables can also have varying amounts of lignin; if there is a great deal of it, we describe them as having a 'woody' mouthfeel. Some vegetables accumulate lignin after they have been harvested, with white asparagus being an especially good example of this effect.

Why Do Asparagus Become Woody?

After they have been cut away from the plant the stems of asparagus continue to be biologically active. Their own enzymes become active and quickly convert the very small amounts of sugar in them into a fibrous substance called lignin, the same material that reinforces the cell wall of plants such as trees. This is why asparagus, especially the white ones, and other vegetables such as broccoli become 'woody' if they are left lying around for any length of time. Fresh asparagus can become tough in as little as twelve to twenty-four hours after they are harvested. It is easy to determine the point between crispness and woodiness on a stalk of asparagus by using the 'snap test.' The less desirable woody parts, just like peelings, can be set aside and cooked to make a soup stock with a delicate asparagus taste. When the sugar is converted to lignin the asparagus will taste less sweet. Lignin formation takes place more quickly in a warm, well-lit environment. It can be slowed down by keeping the cut asparagus in a dark place at a temperature between 2 and 10 °Celsius with a humidity level of 95–100 percent.

Pectin and Crisp Vegetables

Many of the soluble dietary fibres, for example, pectin, are found between the cells in the plant tissue. When pectin molecules are dissolved in water, they become negatively charged and repel each other. In order to bind the pectin molecules tightly to each other and form a firmer structure together with the water, that is, create what we call a gel, it is necessary to counteract this repulsion. One way to do so is to add sugar (sucrose), which binds water and, as a consequence, forces the pectin molecules to link together a little more strongly. Another possibility is to add acid, which reduces the extent of the electric repulsion. Still another option is to introduce calcium ions or magnesium ions, which have a positive charge and, in this way, bind the negatively charged pectin molecules together. Which of these possibilities results in the strongest gel depends on the type of pectin in question.

The ability of calcium ions to stiffen foodstuffs that contain pectin can be exploited to ensure that cooked or pickled vegetables remain firm. It is simply a matter of adding sea salt or a pure calcium salt such as calcium citrate.

It Is All about Reducing Water Content

Preserving vegetables is partly a matter of reducing their water content, but it is not necessary to remove it entirely. What is really important is lowering what is known as the activity of the water. This expression refers to a measure of the extent to which the water molecules in an ingredient are actually accessible in the sense that they are free to participate in other processes. If the water molecules have already formed tight bonds with other molecules, their activity is lower. A food product, for example, dried fish, can easily have a water content of 20 percent, but it can keep for years and is fully preserved because the water is so tightly bound in the tissues of the fish that microorganisms will not be able to survive.

There is a whole range of different dehydration processes that can be used to reduce the activity of water. One of these is to blow dry, possibly warm, air over the raw ingredient. Another is freeze-drying, in which the pressure around an already frozen substance is lowered to the point where the water sublimates, changing directly from its solid state, ice, to water vapour. It is also possible to take advantage of the osmotic effect by using salt or sugar to draw liquid out of the foodstuff. Alcohol works in a similar, but less effective, manner, although the alcohol itself acts as a preservative. Smoking is also somewhat dehydrating. Finally, allowing certain harmless moulds to grow on the surface of a food can help to dry it out. This effect is very visible on many costly air-dried hams, some aged sausages, and specialty fish products such as the Japanese *katsuobushi*.

When making *tsukemono*, which makes use of the osmotic effect to draw water out of the vegetables, it may be necessary to drain away the liquid from time to time. In some cases, this also becomes an important way to remove undesirable, bitter taste substances that have dissolved in the liquid that seeps out.

Even though a fair amount of water is removed by the different techniques for making *tsukemono* described below, this is not sufficient to preserve them. Consequently, their shelf life is completely governed by their contents of salt, sugar, acid, and alcohol. In some cases, preserving agents are also used. Similarly, it is important to pay attention to how long it is safe to keep a particular vegetable pickle before it is consumed. Some commercial products are pasteurized or have added preservatives and are placed in tightly sealed packages to increase keeping qualities. Refrigeration may also be necessary.

Dehydration

Vegetables can be dehydrated very simply by placing them out of doors in sunny, dry weather. In parts of Japan, this was the traditional way to handle vegetables that were to be made into *tsukemono*. For example, *daikon* were often hung under the eaves of the houses.

It is, however, best to dry foodstuffs in a controlled manner in an oven or a dehydrator. The drawback with using an oven is that the temperature can be too high, that is, over 50 °Celsius and the vaporized liquid has no way to escape. The solution is to use a dehydrator that blows warm air over the vegetables and at the same time removes the vaporized liquid as it is formed. This ensures that the temperature on the surface of the vegetables remains constant.

Typically, the temperature in a vegetable dehydrator is kept in the range of 40–60 °Celsius in order to preserve colour, as well as taste and aroma. The partial dehydration of the vegetables diminishes the growth of bacteria and enzymatic activity on their surfaces, thus lessening the chances that proteins will be broken down. Nevertheless, this is not sufficient to preserve the vegetables.

◘ Cucumbers that were cut open lengthwise and had their inner juiciest parts removed before being dehydrated.

The amount of time required to dehydrate a vegetable
differs greatly from one variety to another. As the emis-
sion of water vapour is a diffusion process, one should
also be aware that the drying time depends quadratically
on the thickness of the ingredient. That is to say, a piece
that is twice as thick as another needs to be dried for four

times as long, one that is three times thicker for nine times longer, and so on.

Fresh vegetables typically have a water content of 80–95 percent. Dehydrated vegetables with up to 15 percent water can be kept for a certain length of time, provided they are refrigerated. But the water content generally has to be reduced to under 5 percent in order for dried vegetables to keep without any further conservation. Modern freeze-drying techniques can reduce the water content to as little as 2–3 percent, allowing the vegetables to last for years, provided they are stored in dry conditions. For making *tsukemono* it is not desirable to remove nearly as much water, usually only up to about a half. It is not the dehydration, *per se*, that helps to preserve the pickles, but rather the subsequent marinating in salt and possibly acid.

Dehydration is hard on the cells of the vegetables; it changes their shape and the cell walls can be broken, altering their texture. The sugar content of the dried vegetables also affects their texture. One with more sugar will be chewier and more flexible. Sugar binds the water and will also cause water to be absorbed from the surroundings. Vegetables that have a great deal of soluble dietary fibre, such as carbohydrates like pectin (hydrogels) that bind water less strongly than sugar, have a tendency to turn out crisper than those with a high sugar content.

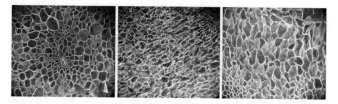

◘ Cell structure of a radish: while fresh (left), after dehydration (middle), and after rehydration (marinating) (right). During dehydration the cells shrink, and the network of the otherwise stiff cell walls becomes crumpled and more flexible. During rehydration the cells again absorb water, but the network of cell walls remains irregular and flexible. The effect of this is that the rehydrated radish is both flexible and has a crunchy texture. The size of the images corresponds to 0.6 mm × 1.6 mm.

Dehydration can change the colour of vegetables, and if the temperature is too high the green colour from chlorophyll can turn yellow or brown. In addition, the colour sometimes appears to have been altered when the surface of the ingredient has taken on a spongy structure that reflects light in such a way that the surface looks duller.

The Pickling Crocks

A variety of containers and pickling crocks can be used when making *tsukemono*; the choice depends on which type is being prepared. But under all circumstances, ensuring that the containers are absolutely clean is paramount as it is otherwise difficult to control the outcome, especially if fermentation is also involved.

Although the raw ingredients are sometimes blanched or scalded on the outside before they are placed in the pickling crocks, heat plays no role in the preparation of *tsukemono*. But as pressure is often involved, the most suitable containers have a drop lid that fits inside the crock or have a pressure-plate. This allows pressure to be applied to the contents, speeding up the pickling process and ensuring that the vegetables are not exposed to air. When the vegetables are also to undergo fermentation, using a pickling crock with a water lock can help to ensure that the process is not spoiled by contact with air.

Miso-zuke, *nuka-zuke*, and *kasu-zuke*, which are all immersed in solid pickling media, are traditionally placed in a wooden container or a barrel (*taru*) that has a lid that will fit inside it. The barrel is made from cypress or cedar wood staves held together with bamboo barrel hoops. A heavy weight, usually a stone, can be placed on top of the lid to put pressure on the contents. A *taru* is cleaned with a special kitchen brush (*tawashi*) made from hemp fibre. The same brush is used to scour away any traces of dirt on the vegetables before they are placed in the barrel. Vegetables such as *daikon* and carrots should not be peeled, but simply scrubbed thoroughly.

◘ Traditional Japanese scouring brush, *tawashi*, for scrubbing vegetables before they are made into *tsukemono*.

For pickling, wooden containers have the advantage that they can absorb and give off liquid, so the moisture level in them is better regulated. They are also able to breathe, allowing the fermentation process to take place more uniformly. But as this brings with it the risk of contamination by undesirable fungi, they require more cleaning and maintenance. Normally this means scalding them with boiling water or rubbing their insides with sake or *shochu*. When space considerations are added to the maintenance work required, the result is that wooden barrels are now rarely used in modern households.

Taru

A *taru* is a special wooden barrel that has been used for centuries in Japanese households to make *tsukemono*. The barrel was made of cypress or cedar wood and was topped by a drop lid that was weighed down by a large, heavy stone. It contained a yellowish fermentation medium, for example, *nuka-doko* based on rice bran, which was inspected and stirred daily. If the barrel was kept in a cool, dark place, the medium could be kept going for years, spanning several generations. A variety of vegetables, such as dried *daikon* (Chinese radishes), were placed in the barrel. After several months they emerged as *takuan-zuke*, tasty pickles with an aroma that the initiated find inviting. It is said that in earlier times Japanese housewives were sometimes called *nuka-miso*, which loosely translates as 'smelly women.' Their characteristic smell was due to their contact with the pungent fermentation medium, *nuka-doko*, which they turned over every day by hand, also causing their hands to take on a yellowish tinge.

■ *Taru*: Classical Japanese wooden barrels used for fermenting *tsuke-mono*.

Tsukemono that are prepared either by using salt or immersing them in a marinade, for example, *shio-zuke* and *shoyu-zuke*, were generally prepared in stoneware or ceramic containers, which are not damaged by salt and the acid that can form during lactic acid fermentation. As with wooden barrels, though, the old-fashioned pickling crock is being replaced by a plastic container (*shokutaku tsukemono ki*) that has a tight-fitting lid. Inside there is a spring-plate combination attached to a screw handle on the lid; this allows the ingredients to be placed under pressure as they marinate. The liquid that seeps out of the vegetables forms a marinade that covers the vegetables. This limits contact with oxygen in the air, which would otherwise facilitate the growth of unwanted bacteria. Some *shokutaku tsukemono ki* will easily fit into a refrigerator and they are hygienic and easy to keep clean.

◘ The modern and the traditional: a modern plastic pickling crock with a movable spring-plate that applies pressure to the vegetables inside and an old-fashioned, ceramic pickling crock for fermentation of vegetables over an extended period of time. The old crock has a water lock to keep out air.

The pickling crocks must not be made of, or contain, any metal as its presence increases the risk of oxidation, which imparts unpleasant aftertastes. The only exception is for metal that has been enameled. Acids, especially citric acid, can partially counteract oxidation by combining with the metal. This is why acids such as ascorbic acid and citric acid are used as antioxidants in foods. Nevertheless, it is best to avoid any contact with metal.

If one does not have a suitable crock and primarily makes quick-pickled *tsukemono* using salt and vinegar it is quite possible to get by with an ordinary plastic container that has a tight-fitting lid, a large glass jar, or a simple plastic bag with a sliding seal.

It is important to pack the vegetables tightly in the container, especially if only salt is involved. This minimizes the risk of contact with the air and ensures that an equal amount of liquid seeps out from all the vegetable pieces and that they are pickled to the same extent. When using semi-solid media such as *nuka* and *sake-kasu* the vegetables should not be packed as tightly, preferably leaving a little space between the individual pieces. The spaces should be filled completely with the pickling medium so that all the surfaces are in contact with it and there are no air pockets.

Brining

Treating the raw ingredients with salt is often the very first step in making *tsukemono*. Salt can draw out up to half of the water content in the vegetables by osmosis. This can be done either by sprinkling the salt on the vegetables, that is, dry-salting, or immersing them in a ready-made brine.

Dry-salting draws out the water gradually, in effect forming a brine. This liquid can be poured away or mixed with a marinade. It is necessary to drain off the liquid if the vegetables are to be placed in a fermentation medium, for example, *sake-kasu* or *miso*.

Salt stimulates the activity of lactic acid bacteria, causing them to produce lactic acid, which also acts as an additional preservative and can suppress the growth of spoilage bacteria. As a result, dry-salting is often associated with fermentation. Acids such as vinegar can, to a certain extent, compensate for a reduced salt content.

The ability of salt to draw liquid out of the vegetables, especially from leaves and green tops, may also allow it to remove bitter substances that could otherwise be eliminated only by boiling the vegetables. In addition, salting may eliminate some of the sharp tastes. Vegetables that are to be marinated or fermented are better able to absorb the nutrients, tastants, and aromatic substances found in these pickling media if they have first been salted.

If salting is to be done over a longer period of time, it is best to place the raw ingredients in a container with a movable lid that is weighted down or that has a pressure-plate in order to exclude air and maximize immersion in the brine. The pressure should be applied carefully, so that the vegetables are not bruised. More pressure needs to be applied in the case of very firm vegetables and fruits. Less pressure is needed if more salt is used.

In principle, there are three, slightly different ways to dry-salt vegetables to prepare *shio-zuke*, namely, *ita-zuri*, *shio-furi, and shio-momi.*

Ita-zuri is an old Japanese technique that is used primarily for preparing whole vegetables such as cucumbers and *daikon*. The salt is rubbed or pressed into the vegetables by hand or by rolling them in salt on a hard cutting board. This serves to soften the vegetables and tenderize their peel as well as ensuring that liquid is drawn out of them. The salt and liquid are then wiped off. More delicate vegetables, for example, cabbage, small aubergines, and vegetable tops can be massaged with salt by hand. They are then squeezed gently to remove salt and water without damaging them unduly.

Shio-furi is a technique used for coarsely chopped vegetables. Salt is drizzled over the vegetables in order for them to 'sweat out' water for a few minutes.

Shio-momi is a method that is especially well suited for use on finely sliced vegetables or the green tops of radishes and turnips. Salt is sprinkled over the vegetables, which are then tossed and may be squeezed a little by hand.

Shio-zuke

Shio-zuke is the easiest and simplest way to prepare *tsuke-mono*. The fresh vegetables are washed and may then be cut up into pieces of a suitable size. They are salted as described above and placed in a clean pickling crock under light pressure to squeeze out the liquid. It is even possible just to shake them together with salt in a plastic bag that seals tightly. Depending on how the pickles are to be eaten, this process can last for a few minutes or hours, or even two or three days. Excess salt can be rinsed or wiped off before the dish is served.

◘ Brining vegetables under pressure.

Shio-zuke can easily be combined with *koji* to ferment the vegetables. The simplest method is to use what is called *shio-koji* (described later), which can be used to marinate a variety of vegetables.

Pickling

The Japanese style of pickling, in the sense that the word is commonly used in other food cultures, can be carried out in two ways—either by using an acidic liquid such as vinegar directly or by introducing microorganisms that produce organic acids in the course of fermentation. Different microorganisms each produce a distinctive organic acid. Two examples are lactic acid produced by lactic acid bacteria and acetic acid produced by the acetic acid bacteria. Generally speaking, the bacteria thrive more readily in their 'own' acid than in those derived from other microorganisms or in acids from external sources. This ability to break down other substances by producing acids is a completely natural consequence of their mutual competition to exist in an optimal environment. Other microorganisms take advantage of waste products. For instance, in the presence of oxygen and an alcohol content that is preferably between 10 and 13 percent, acetic acid bacteria can convert alcohol produced by active yeast to acetic acid. Acids also help to prevent oxidation and inhibit fats from turning rancid.

Pickling is an important aspect of food preservation, as it is possible to suppress the growth of unwanted bacteria by adjusting the acid content. This works in combination with the salt content and the temperature to determine which microorganisms will gain the upper hand during the conservation process. The effect of adding acid to vegetables depends on their buffer capacity, that is, their ability to withstand changes in pH, together with its secondary effect on the ability of the microorganisms to influence the acid balance. As it is not easy to ensure that the raw ingredients are preserved properly using the acidic liquids on their own, salt is almost always required as well. It is also important to be aware that acids have an effect on enzymatic activity. This is why it is not advisable to have too high an acid content when fermentation is being carried out using *koji*, as its enzymes become inactive when the pH falls below 5.

Su-zuke

Su-zuke are made with rice vinegar (*su*). The better-quality Japanese rice vinegar is made from brown rice and has a fuller and more aromatic taste than ordinary household vinegar. A good apple cider vinegar can, however, be used as a substitute. In contrast to the types of pickled vegetables found in Western cuisines, vinegar is an ingredient in only a few types of *tsukemono*.

Marinating in vinegar is almost always combined with salting. If sugar is added instead, the pickles are called *amasu-zuke*. These are often made from *daikon* and onions.

Su

Su is Japanese rice vinegar. Traditionally it is made using either polished brown rice (*komesu*) or unpolished black rice (*kurosu*) according to a very laborious process that involves three different microorganisms. First, *koji*, which is seeded with the fungus *Aspergillus oryzae*, is used to convert the starch in the rice to sugars. Yeast (*Saccharomyces cerevisae*) is then added to ferment the sugars to produce alcohol. Finally, acetic acid bacteria (*Acetobacter pasteurianus*) are introduced to convert the alcohol to acetic acid. *Kurosu* is fermented in stone crocks placed outdoors. It has a greater concentration of free amino acids and other organic acids than other vinegars and, therefore, has a more complex taste with more umami. Even though it is more expensive than other vinegars it has become popular in Japan as a healthy drink. Another type of Japanese rice vinegar is made from sake lees (*sake-kasu*). Japanese rice vinegars have a milder and more aromatic taste than ordinary household vinegar and their acetic acid content is somewhat lower, typically 2–4 percent, compared with 5–7 percent. The quality and complexity of a vinegar's taste depend on how much rice has been used in its production. For the cheapest ones of inferior quality only 40 grams of rice go into each litre, whereas the best ones are made with 300–400 grams of rice per litre. In Japan, rice vinegar is normally used only for quick pickling (*sunomono*) and seasoning, rather than for more involved types of pickling.

Marinating in Soy Sauce, *Miso*, and Sake Lees

Brined *tsukemono* are often transformed into other types by marinating them in yeasted products such as soy sauce (*shoyu*), miso, or sake lees (*sake-kasu*), all of which contribute taste substances, mineral content, and nutritional value. Marinating in *miso* and sake lees is usually carried out over a long period of time.

Both soy sauce and *miso* contain salt and have an abundance of free amino acids, which contribute a strong umami taste. *Sake-kasu* has a great deal of glutamate from the dead yeast cells left over from brewing sake and these are also a good source of umami. In addition, lees have a residual alcohol content that introduces other tastes and helps to preserve the vegetables.

▣ Ingredients for marinating and fermentation media for making *tsukemono*: bran, *miso*, and sake lees.

Shoyu-zuke

Shoyu-zuke are often marinated in a mixture of soy sauce and sake or, possibly, *mirin* if the pickle is to be somewhat sweet. Many different vegetables are prepared this way, resulting in pickles that vary from being very salty and rich in umami to those that are sweeter. The vegetables can be kept in the refrigerator for a very long time, allowing their taste to become more intense as the pickles age. It is quite easy to prepare *shoyu-zuke* in a plastic bag with a vacuum seal.

As will be described later on, a variation on *shoyu-zuke* can be made by adding a mixture of *koji* and soy sauce.

■ *Shoyu-zuke*: marinating white asparagus in soy sauce in a plastic bag with a vacuum seal.

Shoyu

Shoyu is Japanese soy sauce produced according to a process that dates back to 1643. Both soybeans and wheat are used in varying proportions, resulting in different taste nuances. Wheat adds more sweetness and yields a higher alcohol content. The secret behind the brewing process is the fermentation medium, *koji*, without which it is not possible to produce authentic soy sauce. The first step involves seeding cooked soybeans and roasted cracked wheat kernels with the spores of the moulds *Aspergillus oryzae* or *Aspergillus soyae*. After a few days, the mould has grown in the mixture, which is now referred to as *koji*. The *koji* is placed in a brine with 22–25 percent sodium chloride (NaCl) by weight and left for six to eight months. The salt causes the fungus to die, but its enzymes remain active. If fermentation does not start to occur spontaneously, a secondary fermentation culture, consisting of yeast and lactic acid bacteria that will thrive in the salty environment, is added. The action of the residual enzymes and these microorganisms breaks down the proteins and fats of the raw ingredients. This thick mash, called *moromi*, is placed in large, open cedar barrels and allowed to ferment and age for two summers. When this stage is complete, the *moromi* is transferred to linen sacks and pressed to extract the liquid soy sauce and soy oil, which floats on top of the sauce and is drained off. Finally the soy sauce is pasteurized and bottled. For the highest quality soy sauce, the process as described above takes two years. Less than 1 percent of modern Japanese soy sauce is still made using this costly, labour-intensive process. New methods that utilize acids to hydrolyse the soy protein in a matter of days have allowed less expensive products to enter the market, although they are of lower quality and have less taste. Soy sauce contains an abundance of free glutamate, which is a source of umami, and has a very high salt content, on the order of 14–18 percent.

Miso-zuke

Miso-zuke are usually made with *aka-miso* (red *miso*), a dark variety that has a high salt content and is rich in proteins. The paste can be mixed with a little soy sauce, *mirin*, or sake, and possibly seasoned with spices and garlic. This marinade is called a *miso-doko*. The raw vegetables are dried thoroughly and then placed in the mixture, ensuring that they are fully covered with it. A *miso-doko* can be used over and over as long as it is kept in a cool place and is not allowed to become too liquidy.

Vegetables can be marinated for as little as an hour or allowed to age in the *miso-doko* for weeks or months. The taste becomes stronger and more intense with the passage of time. *Miso-zuke* will keep for a long time in the refrigerator. Excess *miso* is wiped off before the pickles are served.

◘ *Miso-zuke* made with garlic.

Miso

Miso is made from fermented soybeans and a variety of grains, such as rice, barley, wheat, buckwheat, rye, and millet. The production of *miso* is closely related to the brewing of *shoyu* and, in a sense, *miso* is the solid mass that is left over when the liquid is drained away from the fermented soybean mixture. As is the case with *shoyu*, the fermentation medium *koji* plays a central role, and yeast and lactic acid bacteria may also be involved. There are many different types of *miso*, with their particular cereal content reflected in their colour, taste, and consistency. The content of rice and barley, as well as the fermentation period, determines the colour of the finished product, which ranges from pale yellowish-white (*shiro-miso*), to yellow or light brown (*shinshu-miso*), to red or dark brown (*aka-miso*), and to very dark *miso*. A typical miso paste has a protein content of 14 percent, as well as large quantities of free amino acids, especially glutamate that imparts umami taste. The salt content ranges from 5–15 percent. It can take up to two years to ferment high quality *miso*. Industrially produced *miso* is prepared over a much shorter time frame and is also sweeter due to its considerable barley content. In addition, taste additives are often mixed into the final product.

Kasu-zuke

Kasu-zuke are fermented in a medium consisting of a mixture of sake lees (*sake-kasu*), sake or *mirin*, salt, and sugar. Traditionally the fermentation process was of long duration and in some cases took several years. The marinade preserves the raw ingredients and improves both their keeping qualities and their nutritional value. Originally these *tsukemono* were made in Japan from pickling melons, aubergines, and cucumbers, but later carrots and ginger root were added to the list.

As *sake-kasu*, sake, and *mirin* all contribute umami to a dish, they can be used to enhance the taste of vegetables that are not very flavourful. Because *sake kasu* also has alcohol and sugar contents of about 8 percent and up to 20 percent, respectively, it helps to impart *kasu-zuke* with their characteristic taste.

Fish can also be cured in *sake-kasu*, as long as the marinade is made without sugar. On the other hand, lemon juice, ginger, and possibly a little neutral tasting oil are good additions. When it is ready, the fish can be eaten as it is, marinade and all, or gently fried on low heat.

ℹ Medium for fermenting *kasu-zuke*

Making a medium for preparing *kasu-zuke* is very easy and takes little time if one is able to obtain sake lees left over from brewing sake.

> 1 kg (2 1/4 lb) *sake-kasu*
> 125 mL (1/2 c) *shochu*, *mirin*, or sake
> 125 mL (1/2 c) water
> Sugar, according to taste
> Salt
> Mix everything together in a bowl, cover it with plastic wrap, and put it in a cool place for a few months. As it ages, the medium will become sweeter and darker. It can also be used right away to make *kazu-zuke*, but the taste of the pickles will not be as intense or complex.

Sake

Sake is a wine made from polished rice and *sake-kasu* is the solid mass left over from brewing it. The highest quality sake is made from rice that has had more than half of each grain polished away. The rice is cooked and fermented using the medium *koji*, which contains enzymes that can break down the starch molecules to sugar and the proteins to free amino acids. Then a yeast culture is added to convert the sugar to alcohol. The free amino acids in the finished sake are derived from the small quantity of proteins left in the polished rice, together with the proteins found in *koji* and the yeast. This is why sake elicits umami taste. The lees, called *sake-kasu* are made up of residual starch and sugar, as well as the spent yeast cells, which contain a great quantity of free amino acids and glutamate that elicits umami taste. As a result, sake lees are very suitable for marinating vegetables and fish.

Fermenting and Yeasting

Our immediate environment is home to a vast number of different microorganisms, which are often harmless. They are also found on our fruits and vegetables and there are significant local variations as to which types are present. This is one of the reasons why the term 'terroir' can be also used when referring to raw ingredients. Under the right conditions, these benign microorganisms can help to hold spoilage and pathological bacteria at bay. In addition, the plants themselves produce substances that partake in the chemical warfare against their dangerous surroundings.

After a plant, for example, a vegetable, is harvested, it becomes a race against time to determine which of the microorganisms will gain the upper hand. In the first instance, when only a very limited amount of oxygen is present, microorganisms such as lactic acid bacteria and yeast convert the readily accessible sugars. This creates a series of by-products including lactic acid, alcohol, and carbon dioxide. These microorganisms have no effect on other substances, for example, certain vitamins and aromatic compounds.

In a sense, we can think of the vegetables as already being seeded with a culture that can initiate fermentation. A possible way of speeding up the process is to add a ready-made, well-established culture as is done when baking sourdough bread. Lactic acid fermentation can be started by adding a freeze-dried starter culture or a salty, active sour milk product such as yogurt.

◘ Pickling crock with Hokkaido pumpkin and cabbage that are undergoing fermentation with lactic acid bacteria.

Which microorganisms will win the race depends very much on ambient conditions such as temperature, salt content, and acidity (pH). Over and above that, the success of a particular microorganism can sometimes lead to its own death, because the substances it has produced when active provide the living conditions for another type of microorganism that then becomes dominant. An example is yeast, which converts sugar to alcohol that, in turn, nourishes acetic acid bacteria, which convert the alcohol to acetic acid.

Similarly, the activity of microorganisms is closely tied to processes involving enzymes that convert one substance to another, which then becomes the growth medium for a different microorganism. Moulds such as *Aspergillus oryzae* (in *koji*), for example, form enzymes that can transform carbohydrates to sugar, which can then be turned into alcohol by yeast cells, a process that takes place during sake brewing.

Control of Salt Content, Temperature, and Access to Oxygen

Controlling the salt content is the single most important aspect of preparing really good *tsukemono*. In conditions of low salinity (under 2 percent), weak acidity (under 0.3 percent), moderate temperature (under 22 °Celsius), and the absence of air, bacteria belonging to the species *Leuconostoc mesenteroides* will thrive. Over time they will produce a mixture of organic acids, alcohol, and a variety of aromatic substances that have a pronounced mild taste. The acids help to suppress the growth of spoilage bacteria. When the salt content is greater (2–4 percent) and the temperature is over 22 °Celsius, the *Lactobacillus plantarum* species are about the only bacteria that will be able to survive. A typical scenario is for the *Leuconostoc mesenteroides* to start the fermentation process and then, as the acid content gradually rises, the conditions become optimal for the lactic acid bateria, which become dominant. At high temperatures, for example, during the summer, the lactic acid fermentation can become so vigorous that the end product takes on a sharp taste. Adding more salt can slow down the process.

There is a risk that undesirable and potentially harmful microorganisms can start to grow if the salinity is very low, the acidity too low, the temperature too high, and there is easy access to oxygen from the air. Under these conditions other types of bacteria may make use of the lactic acid and the medium will lose its acidity. This provides an environment for the growth of spoilage bacteria that create unpleasant taste and aroma substances. These bacteria will also break down the tissue structure of the vegetables, so that they become soft instead of crisp. As all of this implies, the process of fermentation involves a number of complex interactions.

Nuka-zuke

Nuka-zuke are fermented in a medium made with rice bran (*nuka*). A starter culture derived from sake lees or the leftovers from soy sauce or *miso* production is mixed into the bran to initiate the fermentation process. It is important that the conditions for this culture are optimized so that it is not contaminated with other, unwanted microorganisms. The resulting *nuka-doko* is like a living organism and, for this reason, it must be tended to on a daily basis. A portion of it can be set aside to form the basis of a new *nuka-doko*, in exactly the same way as is done with a sourdough starter.

Nuka-doko draws liquid out of the vegetables and this creates good growing conditions for different types of lactic acid bacteria. *Nuka-zuke* have a sour, yeasty smell and taste sour and salty. The vegetables become crisp and crunchy. It is possible to ferment both fresh and dehydrated vegetables in a *nuka-doko*. The best known type of *nuka-zuke* is *takuan*, made with dehydrated *daikon*.

Nuka-zuke absorb vitamin B_1, calcium, and lactic acid bacteria from the medium. To give an example, the vitamin B_1 content of *takuan* is tenfold greater than the amount found in the original *daikon*. It is also possible to augment the mineral content of the *tsukemono* by such means as placing ungalvanized nails in the medium to yield iron, or eggshells to add calcium.

Before the vegetables are placed in the bran medium, they must be massaged with salt according to the ancient technique *ita-zuri* as described earlier. This helps to soften the vegetables and tenderize their peel, as well as ensuring that more liquid is drawn out of them. Small cucumbers and *daikon* can be rolled on a table or board sprinkled with coarse salt or the salt can be rubbed in using the palms of the hands. The extra liquid that is drawn out of them in this way can be bitter and is discarded before the vegetables are placed in the fermentation medium.

The vegetables are added one at a time and packed closely with the bran between them and between the layers. Fermentation time will determine how strong the taste and smell of the finished pickles will be. It is possible to allow the vegetables to ferment for only a few days or leave them in the *nuka-doko* for up to several years.

On account of its considerable fat content, the bran easily goes rancid. So it is advantageous to use a fresh supply of bran for the fermentation medium when making *nuka-zuke*. If a stronger tasting *nuka-doko* is desired, the bran can be roasted lightly before use.

A *nuka-doko* can have a salt content varying from 5–25 percent. In addition, it can be seasoned with mustard (*karashi*), mushrooms, ginger, sake, chili (-*dokotogorashi*) and kelp.

It is also possible to use bran from other types of grain, for example, wheat. The main point is that it acts as a medium that contains nutrients and, in addition, is able to absorb and derive additional nutritional elements (proteins, fats, minerals) from the vegetables with which the *nuka-doko* is 'fed.'

◨ Vegetables prepared in a *nuka-doko*.

If it is not possible to obtain a little of an already active *nuka-doko* to use as a starter, it is necessary first to make one, exactly as is done to make sourdough.

ⓘ *Nuka-doko* starter

1 kg (2 1/4 lb) rice bran (*nuka*)
1 1/2 L (6 1/4 c) liquid (water mixed with beer or sake)
250 g (1 c) salt
A slice of white bread
Vegetables
Optional seasonings such as a little seaweed (*konbu*),
garlic, fresh ginger root, or chili (according to taste)

1. Mix everything together adding the liquid gradually. If necessary, adjust the amount of liquid so that the resulting medium has the consistency of wet sand.
2. Place the mixture in a pickling crock and smooth the surface. Cover the surface with a clean, damp cloth (not with plastic film) to prevent contamination.
3. The fermentation medium is now activated by 'feeding' it with cut-up pieces of vegetables, such as cabbage, the green tops of *daikon*, turnips, or radishes, carrot peelings, and so on. Place about 200 g (1/2 lb) of these vegetable ingredients in a piece of cheese cloth to make it easier to remove them later. It is possible to speed up the process by stirring in a spoonful of organic plain yogurt.
4. The medium is placed in a cool spot and, on a daily basis, the old vegetables are removed and discarded and fresh ones are added. This process is repeated for at least 3 to 4 days, and preferably up to two weeks. At this point, the *nuka-doko* is ready for use.
5. Once the pickling process is under way, the *nuka-doko* must be stirred by hand every day to aerate it. Any excess liquid should be carefully drawn off using a clean cloth or absorbed by adding more bran. It is vital to ensure that everything is kept very clean, for example, by covering the medium with a damp cloth to prevent contamination by undesirable airborne microorganisms. These can ruin both the taste and the nutritional value of the pickles and may even pose a health hazard. As salt is absorbed by the vegetables, more must be added on a running basis to compensate for the amount that is drawn into the vegetables. The pickled vegetables can be removed and replaced with fresh vegetables, in theory over a period of months and possibly even longer.

One reason why there is such a variety of tastes of the traditional home-made types of *nuka-zuke* may be that the different homes have their own naturally occurring bacterial cultures in their surroundings and that these cultures are preserved and strengthened if the same *nuka-doko* is used repeatedly over a period of years. It is said that another reason for the variation from one household to another might be that the persons who tend the fermentation medium on a daily basis, stirring it with their hands, introduce their own personal microorganisms.

Koji-zuke

Koji-zuke are traditional fermented pickles that are based on the Japanese national fungus, *koji* (*Aspergillus oryzae*). Dry grains of rice are seeded with spores of the mould that are then stirred into cooked rice. The mould culture develops in this mixture, which is kept warm for a period of twenty-four hours. It turns into *amazake* ('sweet sake'), which has the consistency of a soft paste. It can be used both to make *koji-zuke* and diluted with water to make a warm drink.

In order to prepare *koji-zuke* it is necessary to allow the mould to develop further and ferment the *amazake*. This will produce a *koji* starter culture, described in the recipe below, that needs only to be mixed with cut-up vegetables to prepare the pickles.

ℹ️ *Koji* **starter culture**

1 1/5 L (5 c) water
500 g (2 1/4 c) short grain rice
300 g (1 1/3 c) *koji*-rice
180 g (3/4 c) salt

1. Cook the rice in water for about 20 minutes, until it is soft and pasty.
2. Cool the boiled rice to a temperature of 60 °C (140 °F).
3. Mix the *koji*-rice thoroughly into the cooked rice.
4. Keep the mixture warm for 24 hours. The result is called *amazake*.
5. Mix the salt into the *amazake* and transfer the mixture into a container with a lid that does not seal tightly.
6. Allow the *amazake* to ferment at room temperature for 2 to 3 weeks. It is now ready for use as a medium for making *koji-zuke*.
7. *Amazake* will keep in the refrigerator for several months.

Unless one is exceptionally dedicated it can be a little daunting and somewhat time consuming to undertake *koji* fermentation from scratch. Luckily, this part of the process of making *koji-zuke* has been greatly simplified by the introduction of a commercially available product that can be used right away.

In 2007, when reading through some old manuscripts about food during the Edo period, Myoho Asari, a ninth-generation member of a Japanese family of *koji* makers, came across the expression *shio-koji* in connection with the use of salt and *koji* for pickling vegetables. This inspired her to create a very simple product, which is a thick liquid mixture of cooked rice, *koji*, and salt. It is, appropriately, called *shio-koji* and is sold in plastic pouches or in glass jars and will keep for years. It is so easy to use that it proved to be a real windfall for Asari-san's old factory. It is not without reason that since then, Myoho Asari has been known as the *Kojiya* Lady.

ℹ️ Shio-koji

300 g (1 1/3 c) *koji*-rice
100 g (2/5 c) fine sea salt
Ca. 400 mL (1 3/4 c) water

1. Mix everything in a bowl so the water just covers the rice.
2. Cover the bowl with a piece of cloth and let it stand at room temperature for 8 to 10 days.
3. Stir the mixture daily.
4. After 8 to 10 days blend the mixture
5. Keep the final *shio-koji* in the refrigerator until use. Keeps for several months.

Shio-koji is a cloudy, grainy thick liquid or paste. For this reason, it is unsuitable for fermenting fish or meat that is to be grilled, because the rice solid turns into a burned crust. It is preferable by far to seek out a version of it called *ikitai shio-koji*, which is less opaque, filtered, flows easily, and contains all the active enzymes found in *koji*.

● *Koji-zuke* made from a variety of finely cut-up vegetables, such as cone cabbage, broccoli, green asparagus, and radishes. *Shio-koji* is distributed over the surface of the vegetables by shaking them in a plastic bag that is inflated and then sealed with the air inside.

Shio-koji can be combined with soy sauce to make a variation of *shoyu-zuke*. This type of mixture, *shoyu-koji*, keeps for up to a year in the refrigerator and can easily be used to marinate a variety of vegetables, mushrooms, and meat. This method combines the tastes of salt and of the fermentation products from the *koji* with the strong umami taste of the soy sauce.

Koji has an almost magical effect on vegetables, so that after only a couple of hours they have taken on a milder, sweeter, and less bitter taste. Their texture is softened, but a certain crispness is preserved, depending on how long the vegetables are left in the marinade. The *koji* imparts a slightly yeasty aroma and a fuller and rounder taste that has significant overtones of umami. Vegetables from the cabbage family, such as broccoli, that have a strong and bitter taste, are made more accessible for those, in particular children, who otherwise tend to reject them.

□ *Koji-zuke* made from cone cabbage, broccoli, green asparagus, and radishes.

Fermented Vegetables in Other Food Cultures

Most food cultures turn to fermentation techniques to preserve virtually every type of raw ingredient—meat, vegetables, fruit, milk, eggs, and grains. Other than Japanese *tsukemono*, the best known fermented vegetables from Asian cuisines are Korean *kimchi* and Chinese *zha cai*. In Western culinary cultures, sauerkraut is the most prominent.

Sauerkraut is made from brined, very finely sliced cabbage leaves that undergo spontaneous fermentation caused by lactic acid bacteria at a high temperature (18–24 °Celsius) over a period of about one to six weeks, depending on how strong it should taste. Normally no seasonings are added, although some recipes call for the addition of caraway seeds. The resulting sauerkraut has an acid content of 1–1.5 percent and a salt content of 1–2 percent, leaving it with a fairly sharp, aromatic and quite salty taste.

Kimchi is prepared from brined vegetables, for example, cabbage, cucumbers, *daikon*, watercress, and seaweed. The entire vegetable is used, including leaves and stalks. Unlike sauerkraut, *kimchi* is highly seasoned with a wide range of other ingredients, including garlic, ginger root, green onions, chili, and fish sauce. As with sauerkraut, fermentation and curing occur spontaneously in the presence of lactic acid bacteria, but at a lower temperature (5–14 °Celsius) and for a shorter period of time, one to four weeks. The vegetables retain their crispness. The acid content of *kimchi* is 0.4–0.8 percent, less than that of sauerkraut, but the salt content of 3 percent is greater. Its smell is very pungent, and the taste sensation is more one of sharpness than sourness, and it is also dominated by the additives, especially garlic and chili.

Zha cai originates in Sichuan and is made from the knobby, tuberous stems of a special mustard plant. The stems are brined, pressed, dried, rubbed with hot red chili paste, and allowed to ferment for up to a year. The finished pickles are very spicy, with sour and salty tastes, but have a very crisp texture. Because of their intense flavours, they are usually cut up into very small pieces and eaten sparingly.

Pickled Cucumbers

Pickled cucumbers merit a section on their own as they are the element in Western cuisines that can most closely be compared to Japanese *tsukemono*.

There are two main types of pickled cucumbers. One is fermented by the action of lactic acid bacteria and the other is really a marinated, quickly prepared dish, used mainly as a side salad.

For fermentation, small, hard cucumbers, both gherkins and pickling cucumbers that are not fully ripened, are chosen. As they have not yet developed a soft interior with many seeds, they will retain their crispness throughout. First they are brined in a 5–8 percent salt solution at room temperature for two weeks. During this time, they undergo spontaneous fermentation due to the lactic acid bacteria found on their skin. When finished, the pickles have a salt content of 2–3 percent and a lactic acid content of 1–1.5 percent. Sometimes the cucumbers are soaked in water to obtain a milder taste, and a little household vinegar and possibly some sugar might be added to them.

The other way of making pickled cucumbers is very quick and proceeds by placing the cucumbers in a marinade of vinegar and salt, so that they end up with an acetic acid content of 0.5 percent and a salt content of 2 percent.

They are then pasteurized and a preservative, sodium benzoate, is added to the pickles. Sodium benzoate kills fungi, yeast, and certain bacteria but is only active if the marinade is sufficiently acidic. The cucumbers are stored with the marinade in tightly-sealed glass jars. When this type of pickle is prepared industrially, the cucumbers sometimes undergo an initial blanching to sterilize their surfaces. Alcohol, sugar, and seasonings can also be added at this point. If none of these measures are taken to preserve the cucumbers, they must be eaten right away.

The French have a special way to prepare small gherkins, which are called *cornichons* and seasoned with tarragon. They are very crisp and crunchy.

◘ Pickled gherkins and *cornichons* in a dill marinade.

Using the various rapid processes to pickle cucumbers results in a less complex blend of taste and aroma substances than is the case with those that are fermented. But their advantage lies in the feasibility of exercising greater control over the texture and salt content of the finished products. It is also possible to help the texture of these 'quick pickles' to remain crisp and crunchy by introducing magnesium or calcium salts, which are naturally occurring ingredients in sea salt. Magnesium and calcium ions cause the pectin in the vegetables to bind more tightly to each other.

Large, fully ripe cucumbers can also be marinated to make a very tasty quickly prepared salad, commonly found in Danish and German cuisines. The cucumbers are sliced very thinly, and then sprinkled with salt and allowed to stand for about ten minutes to draw out excess moisture. In the meanwhile, a sweet and sour marinade is made by warming vinegar and sugar (to taste) together and allowing the mixture to cool. The cucumber slices are

Kosher dill pickles seasoned with garlic are a New York City specialty. They are a traditional accompaniment for smoked meat and pastrami sandwiches. These pickles come in two varieties, either 'full-sour' are completely fermented, or 'half-sour' that have been immersed in the brine for a shorter period of time.

squeezed by hand, placed in the marinade, and allowed to stand for at least an hour. Optionally, seasonings such as ground black pepper, garlic, dill, and mild onion can be added. Because the ripe cucumbers have a very large water content, the cucumber slices will remain somewhat crisp for only about two hours, after which they become soft and limp.

There are a number of methods that will ensure that pickled vegetables will be even crispier, but these involve the use of quite harsh chemical additives such as alum (potassium aluminium sulphate) or slaked lime (calcium hydroxide). Both the aluminium and the calcium ions from these substances help to cross-bind the pectin in the cell walls of the vegetables, which makes them fantastically crisp. As slaked lime is a strong, caustic base, it must be washed off very carefully so that it does not neutralize the acid that is being used to preserve the pickles. It also has the effect of breaking down the proteins in the vegetables, which improves their nutritional value.

Tsukemono in Salads and as Condiments

If you keep a selection of *tsukemono* in the refrigerator you can always impart an exciting, crunchy, and colourful element to a condiment or a green salad. The pickles can also add interest and contrast to a dish such as one consisting, for example, of beans, legumes, and lentils, which often have a mealy and soft texture. By chopping the *tsukemono* into different sizes and shapes and by selecting ones of various colours, even the most boring salad will look appealing and have a pleasing and often surprising mouthfeel.

◨ Small salads of kidney beans (on the left) and avocado (on the right) mixed with different types of *tsukemono*.

Tsukemono for Everyone

In this chapter, we demonstrate that it is quite easy to make some types of *tsukemono* at home using ordinary raw vegetables, fruits, and even flowers that are grown in many parts of the world. Doing so requires little by way of equipment, generally only pickling crocks for fermentation, and a few readily available prepared ingredients such as bran, *miso*, soy sauce, sake, and *mirin*. We also show the reader how to incorporate these pickled foods into simple dishes in ways that showcase their varied tastes and textures.

Some recipes are more elaborate and can best be made with the help of a dehydrator. If a dehydrator is not available, it is possible to use an oven as long as the temperature is accurately controlled. And as a number of them call for sake lees, it would be worth tracking down this ingredient even if this might pose a bit of a challenge.

Our experiments with making Japanese-inspired *tsukemono* using locally sourced raw ingredients resulted both in pleasant surprises and a few disappointments. For example, it turned out that kohlrabi can be made into a surprisingly crisp pickle and that placing raw Jerusalem artichokes in a *miso* marinade results in crunchy, very tasty *miso-zuke*. When fresh garlic is pickled in *miso* it becomes fruity, crisp, and has only a very mild garlic aroma. Orange Hokkaido squash that is fermented in lactic acid becomes extremely crisp and retains its beautiful bright colour. When pickled, common prune plums turned out so well that they can be considered as worthy rivals for the classical Japanese *umeboshi*.

Other attempts to introduce ingredients that were distinctly different from those used for traditional Japanese *tsukemono* were not always successful. Raw potatoes, both with and without their peels, were placed in soy sauce or *miso*, but even after several months they still tasted like raw potatoes with a soapy and starchy aftertaste. Adding a bit of acid made little difference. We were not able to create a variation on *uri nara-zuke* using dehydrated squash instead of Japanese pickling melons and placing them in a marinade of sake lees. The dried squash became very soft and watery. And, rather sadly, green asparagus proved to be quite unsuitable for making *tsukemono*, even though the white ones turned out really well.

It is possible that it is still too early to evaluate the results of our many attempts to make *tsukemono* that will keep for a long time (*furu-zuke*), but we will know more about this in the course of time. It is always difficult to make *tsukemono* that will have excellent keeping qualities when using untested raw ingredients. It is hard to predict how various marinating and fermentation media will change over time and interact with the vegetables. One has to be patient and be prepared for new and unanticipated results, both positive and negative.

In what follows, readers will find instructions for preparing a broad selection of *tsukemono*, as well as suggestions about pairing them with dishes and incorporating them into simple meals. In some cases, the recipes are intended to serve simply as a general guide, without any weights or measures, whereas in other instances they are more detailed. For the latter, quantities are given in both metric and imperial units, bearing in mind that the conversion from the one to the other is somewhat approximate. This is usually not an issue as greater accuracy is normally not needed for the recipes in this volume.

ⓘ Basic *tsukemono* marinade

5 dL (2 c) water
17 1/2 g (3 1/2 tsp) sea salt
10 g (2 tsp) sugar
1 tsp lemon juice
5 g (1/5 oz) *konbu*, broken into pieces

1. Mix together the water, salt, sugar, and lemon juice and blend it or heat it until all the salt and sugar have dissolved. Add the *konbu*.
2. Allow the marinade to cool and store in the refrigerator until used.
3. Season the marinade according to taste, for example, with black peppercorns, cooking sake, wine vinegar, *ponzu*, citrus juice, herbs, and so on.
4. Pour the cold marinade over the vegetables that are to be made into *tsukemono*, covering them completely.

Cucumbers

◘ *Tsukemono* made from dehydrated cucumber marinated in sake and *dashi* with salt, sugar, and *konbu*.

Cucumbers are some of the easiest ingredients to use to make *tsukemono*, both the quickly pickled ones, *asa-zuke*, and the different types that will keep for a longer period of time, *furu-zuke*. All varieties, from the common large slicing cucumbers to small gherkins, are suitable. The main challenge with cucumbers is that they are almost entirely made up of water, which constitutes about 95 percent of their overall weight. Much of the water has to be extracted from the cucumbers in order to make good, crisp *tsukemono*.

Japanese cucumbers are slimmer, firmer, and have smaller seeds than ordinary slicing cucumbers. This is why they are more suitable for pickling as it is not necessary to cut away the seeds and watery inner parts. When using slicing cucumbers to make *asa-zuke*, one first cuts the cucumbers lengthwise and then into 1/2-centimetre sections. This step can be omitted for cucumbers that have many fewer seeds such as Lebanese (or Persian) cucumbers or gherkins, which are very firm; they need only to be cut up. The pieces are placed in a plastic bag and tossed with a little salt to which some seasonings can be added. After a few hours in the refrigerator they are ready. The whole process can be accelerated by rubbing salt into the cucumbers, following the *ita-zuri* technique.

◻ Cucumbers in *sake-kazu*. On the left is a Japanese cucumber that has been marinated in *sake-kazu* for at least half a year. On the right are gherkins that have been dehydrated and then placed in a freshly made *sake-kazu* for about two months.

The traditional northern European cucumber salad can be thought of as a variety of *asa-zuke* with sugar, *amazu-zuke*. But this type of salad does not stay crisp for very long. The problem is that even if some water is extracted by dry-salting the cucumber slices before they are placed in the slightly sweet vinegar marinade, the water content is still too great to allow them to stay crisp for a longer period of time. The solution is to dehydrate the cucumbers so that they lose up to 75 percent of their weight before they are marinated. The Japanese-inspired cucumber salad below becomes very crisp and crunchy and stays that way for several weeks under refrigeration.

ℹ Cucumber *tsukemono*

5 slicing cucumbers (ca. 1600 g or 3 1/2 lb)
10 g (2 tsp) sea salt
5 g (1 tsp) sugar
80 g (1/3 cup) cooking sake
40 g (2 2/3 Tbsp) water or *dashi*
Yuzu juice or lemon juice
5 g (1/5 oz) *konbu* broken into small pieces

1. Slice the cucumbers lengthwise. Use a spoon to scoop out the seeds and inner, juiciest flesh. If using Lebanese or Persian cucumbers, omit this step.
2. Place the cucumbers in a dehydrator at 40–50 °C (104–122 °F) for 10 to 12 hours. Their combined weight should be reduced to about 400 grams (14 ounces).

3. Cut the cucumbers into slices 2–4 mm (1/8–1/4 in) thick. Pack them tightly with the pieces of *konbu* in a plastic container with a tight-fitting lid.
4. Make a marinade with the salt, sugar, water, sake, and a few drops of *yuzu* juice.
5. Pour the marinade over the cucumber slices, squeezing out any air bubbles and making sure the top layer is completely covered. If necessary, add a little more water or sake.
6. Refrigerate. The finished pickles are ready to eat after a few days. They will keep for about two months in the refrigerator.
7. A more intense taste can be achieved by substituting *dashi* for the water.

Dehydrated gherkins are an excellent choice for making *nara-zuke*. They should be kept whole and placed in a dehydrator at 40–50 °Celsius for 5 to 10 hours, then covered with *sake-kasu* and put in a sealed plastic bag and refrigerated. Then it is only a question of exhibiting patience and waiting for several months. Their texture improves and they become darker and more aromatic over time. One can easily remove a few pieces to sample along the way, leaving the rest to continue marinating.

Asparagus

In the spring, white asparagus are an eagerly awaited culinary delight. Even the thickest and the newest stems can be marvellously crisp. They can easily be eaten raw or cooked by steaming them for only a very short time. Sadly, the season for white asparagus is short-lived and preserving and pickling them tends to rob them of their crisp texture, even though this can also result in a much more concentrated taste. Asparagus are rich in umami taste substances.

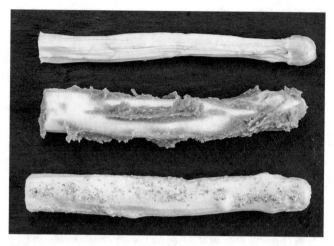

◘ *Tsukemono* made from white asparagus: dried and marinated (top), fresh stalks immersed in *miso* (middle), and lactic acid fermented in rye bread sourdough (bottom).

It is possible to preserve some of the crispness of the white asparagus for many months after the end of their season using a number of techniques for making *tsukemono*. Options include dehydrating before marinating in *miso*, *nuka-doko*, or *sake-kasu*, and lactic acid fermentation. A special method of fermenting raw, fresh white asparagus is to place the stalks in a regular sourdough. This imparts an interesting salty and slightly acidy taste and preserves some of their crispness.

Like a number of other vegetables, white asparagus can also be preserved using *koji*. The peeled stalks are shaken in a sealed plastic bag together with a little *shio-koji*. They remain crisp for a few days and acquire a pleasing aromatic taste.

ⓘ *Koji* marinated white asparagus coated with grated Jerusalem artichoke *miso-zuke* and dried mushrooms

White asparagus

Koji

Jerusalem artichoke *miso-zuke* (see recipe on following page)

Dried mushrooms

1. Peel the white asparagus, shake them with *koji* in a plastic bag and refrigerate them for 24 hours.
2. Grate the Jerusalem artichoke *miso-zuke* onto a plate.
3. Crush or chop the dried mushrooms and mix with the grated Jerusalem artichoke.
4. Leave a little of the *koji* on the asparagus stalks and roll them carefully in the mushroom and artichoke mixture.

◻ *Koji* marinated white asparagus coated with grated Jerusalem artichoke *miso-zuke* and dried mushrooms.

Jerusalem Artichokes

Even though peeling Jerusalem artichokes can be very trying, it is worth the effort if they are to be made into *miso-zuke*. The raw, peeled Jerusalem artichokes are covered completely with *miso* and refrigerated. In the course of two or three months, this variety of *miso-zuke* develops into an exceptionally crispy and crunchy pickle. The vegetables take on a golden, beautiful brown colour, which adds visual appeal.

Jerusalem artichoke *miso-zuke* can be eaten on their own as a little side dish or as part of a snack or an appetizer. When grated, these fantastically crisp pickles are also an ideal way to add taste and texture to other dishes, especially those with fish or poached eggs.

ⓘ Poached eggs with Jerusalem artichoke *miso-zuke* slices and grated oysters

Fresh oysters
Jerusalem artichoke *miso-zuke*
Neutral tasting oil
Eggs, one per serving

1. Stretch a piece of plastic wrap across a deep plate. Remove the oysters from their shells, place them on the film, and freeze them until frozen through. This way they are easy to use.
2. Using a truffle slicer or a mandoline, slice the Jerusalem artichoke *miso-zuke*.

3. Line a small bowl with plastic wrap, allowing it to drape over the sides, and coat it with the oil.
4. Crack an egg open into the bowl and season with salt and pepper.
5. Draw the sides of the plastic wrap up over the egg and twist it to seal. It may be tied with a piece of string to ensure that the wrap does not open up during cooking. Repeat this step with the remaining eggs.
6. Poach the eggs for about 3 1/2 minutes for medium to large eggs. Remove the wrap and place the eggs on a plate.
7. Place the *miso*-zuke slices on top of the eggs. Grate the frozen oysters over them just before serving.
8. Serve at once.

◘ From left to right: Fresh Jerusalem artichoke, Jerusalem artichoke *nuka-zuke*, and Jerusalem artichoke *miso-zuke* (without and with peel).

◙ Poached eggs with Jerusalem artichoke *miso-zuke* slices and grated oysters.

ⓘ Crisp fried rice paper with seaweed salt and grated Jerusalem artichoke *miso-zuke*

Edible rice paper

Neutral tasting oil

Dried seaweed, e.g., dulse or green string lettuce

Maldon sea salt

Jerusalem artichoke *miso-zuke*

1. Fry strips of rice paper in neutral tasting oil at 165 °C (330 °F) until they puff up. Place them on a piece of paper towel to absorb excess oil.
2. Select a variety of seaweed pieces. (If they are fresh, they must be dehydrated first.) Crush the pieces lightly and mix them with the salt in the proportion of 2 parts seaweeds to 1 part salt.
3. Sprinkle the seaweed salt on the rice paper crisps and grate Jerusalem artichoke *miso-zuke* over them.

◘ Crisp fried rice paper with seaweed salt and grated Jerusalem artichoke *miso-zuke*.

Broccoli

Broccoli, like other members of the cabbage family, has a characteristic bitter taste and gives off a distinctly sulphur-like odour when cooked. As a result, even though it is a very nutritious vegetable, broccoli has, somewhat undeservedly, gained a bad reputation. Unfortunately, many people, especially children, have linked its taste and smell to its potentially beneficial effect along the lines of "if it's good for us, it cannot taste good," which gives them an excuse to reject it. But we are willing to bet that the majority of these same people, including the children, would find broccoli in the form of *tsukemono* much more appealing.

One way to do so is to make it into *koji-zuke* by shaking cut-up florets and finely shaved stems in a sealed bag with a little *shio-koji*. After only a few hours, some of the bitterness will have disappeared and the broccoli will have taken on a sweeter, more aromatic taste. And, as it has not been heated, there is no 'rotten egg' smell. This is also true for ordinary white cabbage, in just the same way as radishes and *daikon* taste milder after they have been exposed to the enzymes in the *koji*.

ⓘ Broccolini in *koji* with chili sauce

Broccolini
Koji
Lime juice
Fish sauce
A little sugar
Chili peppers

1. Trim the broccolini and place them in a sealed plastic bag with the *koji* for at least 24 hours.
2. Mix together equal parts of lime juice, fish sauce, and sugar. Stir in seeded, finely chopped chili peppers, season to taste, and refrigerate.
3. Arrange the broccolini on a plate and drizzle the chili sauce on top.

◘ Broccolini in *koji* with chili sauce.

Kohlrabi

Dehydrated kohlrabi are particularly well suited for preparing *furu-zuke*. For best results, select smaller kohlrabi early in the season, before their interiors have become too woody and porous. Their white colour and glass-like characteristics are preserved after marinating. The tops of baby kohlrabi can also be immersed in the marinade to create a colour contrast to the white root vegetable.

ℹ️ Kohlrabi *tsukemono*

4 medium sized kohlrabi (ca. 1200 g or 2 2/3 lb)

5 g (1/5 oz) *konbu* broken into small pieces

10 g (2 tsp) sea salt

5 g (1 tsp) sugar

80 g (1/3 cup) cooking sake

40 g (2 2/3 Tbsp) water or *dashi*

Yuzu juice or lemon juice

1. Cut the kohlrabi up into thirds and place them in a dehydrator at 40–50 °C (104–122 °F) for 12 to 24 hours, according to their size.
2. Trim away the driest edges and the peel, if it appears to be tough. The total weight should now be 250–300 g (1/2–2/3 lb).
3. Cut the kohlrabi into thin slices 2–3 mm (1/8 in) thick. Pack them tightly with the pieces of *konbu* in a plastic container with a tight-fitting lid.
4. Make a marinade with the salt, sugar, water, sake, and a few drops of *yuzu* juice.
5. Pour the marinade over the kohlrabi slices, squeezing out any air bubbles and making sure the top

layer is completely covered. If necessary, add a little more water or sake.

6. Refrigerate. The finished *tsukemono* are ready to eat after a few days. They will keep for about two months in the refrigerator.

A more intense taste can be achieved by substituting *dashi* for the water.

▣ *Tsukemono* made from dehydrated kohlrabi in a marinade of sake and *dashi* with salt, sugar, and *konbu*.

Daikon, Carrots, and 'Vegetable Pasta'

Daikon, which are now quite commonly available, are a real bargain, being one of the cheapest vegetables in terms of cost per unit of weight. In Japan, they are probably the most iconic and classical ingredients for making *tsuke-mono* of all types. One of the main reasons is that they can be made into exceptionally crunchy pickles that keep well and are very versatile. So, we think it is time to bring *daikon* into the spotlight and present them to a wider audience.

◧ *Tsukemono* made from dehydrated *daikon* in a marinade of sake and *dashi* with salt, sugar, and *konbu*.

The best known type of pickle made with *daikon* is *takuan-zuke*, consisting of the dehydrated vegetables placed in a *nuka-doko*. It is important not to let the *daikon* dry out too much as they will become too hard and less crisp. In general, control of the moisture content is of major importance when making *takuan-zuke*. This is why they are traditionally made in wooden barrels, which help to regulate the amount of liquid in the mixture. But there is no problem with using a ceramic pickling crock as long as one keeps an eye on the *nuka-doko* to make sure that it is neither too moist nor too dry.

Daikon are made into *furu-zuke* in exactly the same way as kohlrabi. If they are very small (with a diameter of up to 3 cm) they should be dehydrated whole, while thicker ones should first be cut in half lengthwise. It is normally not necessary to peel the *daikon* before or after dehydrating them.

Like *daikon*, carrots that have first been dehydrated can be used to make *tsukemono*.

■ *Daikon* and carrot *tsukemono* prepared to resemble spaghetti and fettucine.

It is possible to capitalize on the ability of *daikon* and carrots to retain their crispness after being marinated to prepare a sort of vegetable spaghetti and fettucine. The *daikon* are cut into thin strips by hand or using a spiralizer and then dried. Unless the strands are laid out individually it is impossible to prevent them from clumping together as they lose water, but this is not particularly important. It is vital, however, to ensure that they do not dehydrate so completely that they break when they are placed in the marinade. Here they will absorb liquid and taste substances and can then be used to prepare a cold dish of uncooked 'vegetable pasta.'

🛈 *Daikon tsukemono* fettucine with seaweed and grated dried duck heart

1 1/2 g (1/3 tsp) sea salt
1/4 dL (5 tsp) soy sauce
1 Tbsp sugar

Pinch of cayenne

4 duck (or chicken) hearts, ca. 150 g (5 oz)

Seaweeds, preferably green string lettuce or fresh new shoots of bladder wrack

Daikon tsukemono cut into strips like fettucine

1. Make a marinade from the salt, soy sauce, sugar, and cayenne.
2. Trim the duck hearts, rinse away any blood, and place them in a sealed plastic bag with the marinade for 4 hours.
3. Dry off the hearts and place them in a dehydrator or an oven set at 70 °C (160 °F) for 2 hours.
4. Blanch the seaweed in boiling water and immediately plunge it into an ice water bath.
5. Mix the *daikon* fettucine with the seaweed. Grate the dried duck hearts over the dish.

◘ *Daikon tsukemono* fettucine with green string lettuce and grated dried duck heart.

Radishes and Turnips

Turnips, especially the very small ones, and radishes, of all shapes and colours, can easily be made into *tsukemono*. Black radishes have a special esthetic appeal as their dark skin makes an interesting contrast to their white interior when they are sliced cross-wise, but white and red radishes are equally desirable for preparing *furu-zuke*.

In Japan, there is a highly regarded type of *tsukemono* called *nozawana*, meaning 'green things from Nozawa.' These pickles, which are made from the green stems and leaves of a particular variety of small, round turnip, have an unusual history. In the 1700s, the turnip plants, which were native to the mountains around Kyoto, were brought by a monk to Nozawa in the Nagano Prefecture. Here, the climate is much cooler and the vegetables took on different characteristics. Their taproots were considerably smaller,

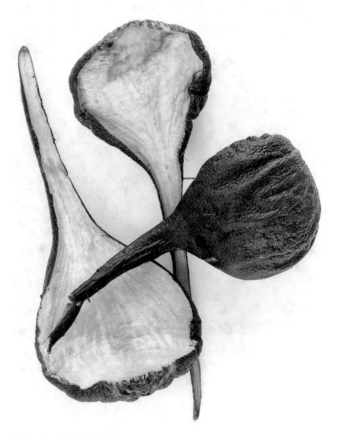

◨ Smoked black radish.

but their stems grew longer and they developed larger leaves that were both soft and succulent. Traditionally the pickles are prepared in Nozawa on one day in December by washing the leaves in the sulphur-laden hot springs found in the area. This imparts a distinctive, characteristic taste to the vegetables. They are then salted and allowed to ferment. Dried sardines or *katsuobushi* flakes are sometimes packed in among the stems and leaves. After these *nozawana* have marinated for three weeks they have acquired a slightly crunchy mouthfeel and an unusual, but pleasing, aftertaste.

◘ Radish and Jerusalem artichoke *tsukemono* served with smoked herring and raw egg yolk.

Turnips and radishes can be made into *furu-zuke* or marinated using exactly the same methods as those already outlined above for kohlrabi. It is also possible to prepare a quick *shio asa-zuke* using finely sliced small turnips that have been rubbed with salt and refrigerated for two hours.

Senmai-zuke can be prepared from thin slices of small turnips or radishes that are marinated together with *konbu* or another type of seaweed, such as dulse, in sweet rice wine mixed with a little salt. Because of the polysaccharides that are exuded by the seaweed, the resulting *tsukemono* have a slightly slimy mouthfeel, but nevertheless they are very crisp and have a powerful umami taste.

One of our experiments involved smoking large black radishes in order to prepare *tsukemono* that would resemble a special, regional Japanese pickle, *iburi-gakko*, made from dehydrated, smoked *daikon*. The radishes were sliced open lengthwise, cold smoked for two or more days, and then placed in the same *furu-zuke* marinade as that used for kohlrabi, *daikon*, and turnips. This imitation *iburi-gakko* is a wonderful condiment for a simple fish dish, for example, smoked salmon or marinated mackerel.

◘ *Tsukemono* made from black radishes marinated with sake, *dashi*, salt, sugar, and *konbu*.

ⓘ Marinated mackerel with smoked radish *tsukemono* and *sansho* peppercorns pickled in *ponzu*

Mackerel

Sea salt

Yuzu

Mirin

Soy sauce

Ponzu

Sansho peppercorns, pickled in soy sauce

Radish *tsukemono*

1. Cut a fresh mackerel up into filets, sprinkle with a little sea salt, and refrigerate for an hour.
2. Rinse off the filets, dry them, and place them in a *yuzu* and *mirin* marinade for an hour.
3. Mix together equal amounts of soy sauce and *ponzu* with the *sansho* peppercorns and cook gently until the sauce thickens. Allow it to cool.
4. Take the filets from the marinade and remove the outer membrane of the skin and the small bones.
5. Cut the mackerel filets into long strips, slice the radishes into thin discs, and arrange on plates. Drizzle with the sauce and serve.

■ Marinated mackerel with smoked radish *tsukemono* and *sansho* peppercorns pickled in *ponzu*.

When making *tsukemono* from radishes, choose ones that are not too old or porous. Although they often have a very sharp taste, pickling can take the edge off it without eliminating it entirely. This can be done by using the method already described for preparing *koji-zuke* from broccoli, by dehydrating them and following a recipe for making *furu-zuke*, or by lactic acid fermentation. Long red radishes give the best result and it is possible to preserve their beautiful purplish red colour.

🔲 Fresh radishes (top) and dehydrated ones that have been lactic acid fermented (bottom).

◻ A new take on relish. Serving of scampi, *miso*-mayo, and two types of *tsukemono*. On the left, whole pieces of different types of *tsukemono*. On the right, a relish made from a variety of coarsely chopped *tsukemono* mixed with *miso*-mayo.

Chinese Cabbage and Lacinato Kale

The culinary potential of Chinese cabbage is perhaps undervalued because it has often been used, without much thought, as a coarsely chopped addition to an ordinary salad. But it is actually fantastically well suited for making a tasty, interesting, and impressive version of *tsukemono* that is similar to the famous Japanese *hakusai-zuke*. Here we try to give it the prominence it deserves.

It is very simple to turn Chinese cabbage into a type of *shio-zuke*. The head is left whole and rubbed with salt, making sure that some of the salt goes down between the leaves. It is then squeezed together, tied with a string, and refrigerated, at which point it spontaneously starts to undergo a weak lactic acid fermentation process. As 95 percent of the cabbage is made up of water, much of it is extracted by the salt, but the vegetable retains its crispness. After a few days it is ready to be eaten, possibly dressed with a marinade containing spices such as chili.

Black lacinato kale, which is sometimes called dinosaur kale and is from the same plant family as the more familiar green curly kale, has enjoyed a renaissance in the last few years. It is hardy and has the advantage of being available fresh far into the winter season. Making lacinato kale into *koji-zuke* imparts it with a desirable umami taste.

■ *Koji-zuke* made from lacinato kale.

■ A head of Chinese cabbage that has been rubbed with salt and has undergone a weak lactic acid fermentation process.

Garlic

Both the taste and texture of garlic change quite considerably depending on how it is prepared. When it is cooked, its sharp, characteristic aroma and taste almost disappear, but many of these subtle aspects can be preserved when it is prepared without heat, a typical example being black garlic. Similarly, using fresh, peeled garlic cloves to make *shoyu-zuke* or *miso-zuke* brings out many surprising taste and textural elements in them. The latter can be sliced into thin flakes to accompany fish dishes.

ⓘ Salmon *tataki* with crushed soy sauce coated pumpkin seeds, Chinese cabbage *shio-zuke*, and garlic *miso-zuke*

Neutral tasting oil

Pumpkin seeds

Soy sauce

Large skinless salmon filet cut from the centre of the fish

Chinese cabbage *shio-zuke*

Flakes of garlic *miso-zuke*

Lemon

1. Pour a small amount of oil into a skillet and toast the pumpkin seeds over medium heat. Allow them to puff up a bit and then add a little soy sauce. Continue heating to reduce the soy sauce.
2. Allow the seeds to cool off on paper towel. Then crush them with a mortar and pestle.
3. Season the salmon, sear it, and coat it with the crushed pumpkin seeds.
4. Take a whole leaf of the pickled Chinese cabbage, press it a bit to remove some of the moisture, and arrange on a plate. Place slices of salmon beside it and sprinkle them with the garlic flakes.
5. Serve with a wedge of lemon.

◘ Salmon *tataki* with Chinese cabbage *shio-zuke* and flakes of garlic *miso-zuke*.

Squash

When raw, orange Hokkaido squash or Hubbard squash are far too hard to be edible, even if they have been peeled. But they can undergo a transformation if they are subjected to lactic acid fermentation, for example, in a crock together with cabbage and other vegetables. It typically takes several weeks or a few months for this process to yield the best results. The squash become very crisp, can be eaten with their peel, and retain their wonderful orange colour. Another type of firm squash, Hubbard squash, also has pickling potential.

◘ Orange Hokkaido squash and white cabbage that have undergone lactic acid fermentation.

◘ Fresh goat cheese topped with fermented orange Hokkaido squash and dried olives on toasted sourdough bread.

Ginger Root

Just about everyone who has ever eaten sushi is familiar with one kind of *amazu-zuke*, namely, sweet and sour pickled ginger root, known in Japanese as *gari*. The versions of this condiment served in restaurants are almost always commercial products, which are often coloured with red *shiso* or food colouring to turn them pale pink. In addition, *gari* contains citric acid, acetic acid, and preservatives. But it is quite easy to prepare it at home using fresh ginger root.

ⓘ Pickled ginger root (*gari*)

500 g (1 1/8 lb) fresh ginger root
1 Tbsp salt
375 mL (1 1/2 c) rice vinegar
200 g (7/8 c) sugar
Red *shiso*, optional

1. Peel the ginger roots and slice them very thinly using a mandoline. Try to follow the particular contours of the pieces. It is important to work quickly to prevent the ginger slices from oxidizing and turning brown.
2. Place the ginger in a bowl, sprinkle the salt on top, and rub it thoroughly into the slices using your hands.
3. Refrigerate the ginger for about an hour.
4. Drain away the liquid that has been drawn out of the ginger. Wrap the slices in a clean dish towel or paper towels and press gently to squeeze out some of the remaining liquid. Transfer the slices to a bowl.
5. Heat the rice vinegar and sugar in a saucepan, stirring constantly, until all the sugar has dissolved, and the mixture has reached the boiling point.
6. Pour the marinade over the ginger.
7. Transfer the ginger and the marinade to a jar that has been scalded. Refrigerate. The finished *gari* is ready to eat after two or three days and it will keep for several months.
8. If you wish to add some colour to the *tsukemono*, you can place some red *shiso* leaves or dried, salted ones (*yukari*) in the jar.

◘ Pickled ginger root (*gari*) coloured with red *shiso*.

Danish Open-Faced Sandwiches Made with *Tsukemono*

The open-faced sandwiches that form part of a traditional Danish smorgasbord typically incorporate fresh, ready-made, and preserved ingredients. Fish, shellfish, meat, eggs, cheese, and various sorts of vegetables and herbs all figure prominently. They are combined to showcase their varied textures, colours, and shapes, and then arranged artfully on thin slices of buttered dark rye bread. On occasion these sandwiches may find their way into the lunch box.

Tsukemono give rise to the possibility of re-imagining these open-faced sandwiches so that they are strictly vegetarian and even appeal to children. We tried this out at an agricultural fair where we invited children to put on chef's aprons and attend a cooking school in a tent for an hour and a half. By the end of the lesson, they had made their own bag lunch, filled with vegetarian sandwiches loaded with pickled vegetables. When the children emerged from the tent, they were beaming with pride as they happily showed their impressive culinary creations to their siblings and parents.

Three of the most traditional Danish open-faced sandwiches are made on thin slices of dark rye bread topped with:

- smoked roast pork, 'Italian salad' made with peas and carrots mixed with mayonnaise, garnished with cucumber and tomato;
- breaded fish filet, remoulade, dill, lemon;
- Danish style liver paste with pickled beets, beef aspic, and cress.

As shown in the photo, these served as the inspiration for interesting imitations using pickled vegetables:

- slices of smoked kohlrabi *tsukemono*, salad of white asparagus, dried unripe strawberries, and *tsukemono* made from carrots, horseradish, and celery that has been immersed in bran *doko* overnight, decorated with cucumber *tsukemono*;
- *tsukemono* made from dehydrated and marinated squash that has been breaded and fried, remoulade of different types of *tsukemono*, decorated with a lemon slice;
- smoked fresh soft cheese, *tsukemono* made from dehydrated beets marinated in salt with *konbu*, topped with onions, cucumber juice aspic, and cress.

◘ Danish open-faced sandwiches made with *tsukemono*.

Plums

Japanese *umeboshi* are among the oldest and most classical varieties of *tsukemono*, with a history that goes back at least a thousand years. They are a type of *shio-zuke* that can also be made with the addition of plum wine and marinated. In Japan, these are prepared from several cultivars of a plum-like apricot called *ume* (*Prunus mume*).

The term for this pickle is derived from *ume*, the name of the fruits, which have been sun-dried for three days, and *boshi*, meaning 'something that is dried.' In earlier times, *umeboshi* were included in the field rations for Japanese soldiers, partly to ensure that their muscles were replenished with sufficient salt after hard physical exertion and partly to stimulate the appetite. Today, they are often served in restaurants as a component in a *bento* box and are tucked into bag lunches. Over time, *umeboshi* have taken on a milder taste that is less salty and sour. But they are still one of the most popular forms of *tsukemono* in Japan.

◘ From left to right: Salt-pickled prune plums coloured with red *shiso*, large *umeboshi*, small *umeboshi*.

It is certainly possible to make a version of *umeboshi* using a number of plums that are readily available where stone fruits are cultivated. It is important to choose those with a skin that is not too thin or fragile so that they do not break open while being processed or after they have been stored for a time. Ordinary prune plums, which are quite firm, are good candidates for pickling. So are the smaller greengage plums, although they are more delicate and keep for a shorter time.

Umeboshi can be served as appetizers, eaten with plain cooked rice, infused to make a tea, or used as the filling in a *maki*-sushi roll.

ℹ Prune plum *umeboshi*

1 kg (2 1/4 lb) prune plums
100–150 g (2/5–2/3 c) salt
Red *shiso* or *yukari*

1. Wash the plums carefully and place them on a wire rack for three days.
2. Arrange the plums in a pickling crock with a lid that can support a heavy weight. It is very important to maintain sterile conditions during the pickling process and to scald the crock before using it.
3. Sprinkle the salt on the plums and let them stand for 2 days, turning them carefully a few times.
4. Place a weight on the lid and leave the plums for 4 to 5 days.
5. Add the red *shiso*, which imparts taste and colour and also acts as a preservative. If no dried or salted *shiso* leaves are available, one can substitute 2 Tbsp of *yukari*, a granular mixture of dried red *shiso* and salt.
6. Let the plums marinate in the liquid that has been drawn out for about a week. At this point, traditionally made *umeboshi* are usually dried off and stored in jars. It is also possible to leave them in the liquid and refrigerate them.

◻ *Umeboshi maki-sushi.*

Flowers

a lovely spring night
suddenly vanished while
we
viewed cherry blossoms
Matsuo Bashō
(1644–1694)

Fresh flowers can be preserved simply by shaking them thoroughly with salt. In Japan, the cherry blossom season, *hanami*, is one of the high points of the yearly calendar, a time when family and friends gather to sit under the cherry trees and admire the ephemeral beauty of the flowers. Their enjoyment is tinged with nostalgia and the realization that the blossoms (*sakura*) last for such a short time.

But there is a way to preserve the memory of the flowers and their aromatic scent for the rest of the year. The answer is to salt-pickle blossoms that are not yet fully open to make *sakura no hana no shio-zuke*, which can last for a very long time. The pickles can be steeped in freshly boiled water to make a type of tea. This is very aromatic but, naturally, somewhat salty.

An even more intense cherry taste can be obtained from the cherry tree leaves, which can also be salt-pickled and preserved for a long time in the brine. A traditional Japanese sweet confection (*sakura-mochi*) consists of a ball of sweet, glutinous rice wrapped in a salt-pickled cherry leaf. Even if one does not want to eat the leaf, one will still experience a cherry taste as it seeps into the rice.

ⓘ Salt-pickled cherry blossoms

The best blossoms come from Japanese cherry trees as they have large, very full petals, but the flowers from ordinary cherry trees, both the sour and sweet varieties, are also suitable. The flowers should be picked before they are fully open.

200 g (7 oz) fresh cherry blossoms
3 Tbsp salt
2 Tbsp plum wine vinegar (*umezu*)

1. Gently wash the blossoms, shake off the excess water, cover them with salt, and place them in the refrigerator for 2 days in a container with a lid that can press down on the flowers.
2. Drain off the liquid that has been drawn out by the salt, add the vinegar, and leave the container in the refrigerator for 3 days.
3. Drain off the liquid one more time and spread the flowers out to dry for 3 days at room temperature.
4. Place the flowers in a tightly sealed jar with more salt and shake it so that the salt is distributed.
5. The flowers will keep under refrigeration for several months.

◻ *Sakura no hana no shio-zuke*, salt-pickled cherry blossoms.

Tsukemono in Japan

More than a million tons of *tsukemono* are produced in Japan every year. Each area has its own local specialties and production is geared to the seasons of the year, particularly when it comes to *asa-zuke*, which keep for only a short time. Many families in the countryside still prepare their own pickles and there are a large number of small enterprises that sell their products locally. But by far the greater proportion of *tsukemono* is now made in special factories.

'Preserving the Japanese Way'

This is the title of Nancy Singleton Hachisu's 2015 book about traditional Japanese methods for preserving and fermenting food and how these can be applied in the modern kitchen. She expands on this idea in the introductory section entitled 'Preserving a way of life,' a clear indication that her intention is to deliver a message that goes far beyond what we would normally expect to find in an ordinary cookbook.

Nancy Hachisu's point of departure is her own story, that of an American woman who married a Japanese farmer and for more than thirty years has made a home on his family's farm in the Japanese countryside. Here she observed the extent to which many Japanese had become disconnected from their original food culture and no longer knew how to prepare traditional dishes. Some could barely even identify the ingredients that went into them.

Preserving the Japanese Way together with Hachisu's earlier book about Japanese country cooking, *Japanese Farm Food*, constitutes her openly stated goal of embracing for herself a food culture that is in danger of disappearing and then sharing it with others in all humility. This is partly a matter of convincing them, as stated in many of the recipes in the book, that this way of preparing food is easier than one thinks.

These books are so convincing because they are based on her personal experiences. They tell the story of her arrival in Japan as a newly-wed who had to adapt to a completely new environment and to figure out how to live and think as the Japanese do. Her path forward was driven by a passionate desire to learn to do things in the old-fashioned manner in a rural household.

The author describes how, much against the will of her father-in-law, she gradually wrested control of the kitchen from her mother-in-law and husband, who insisted on preparing the Japanese dishes. It was a matter of plunging in and sticking to what she calls the four mystical elements of rural Japanese cuisine: (1) to slice finely with sharp knives, (2) to roast and grind sesame seeds in a mortar, (3) to prepare the soup stock *dashi*, and (4) to roll hand-made *udon* noodles. All of these culinary arts call for time and patience, commodities that are in scarce supply in modern society. Furthermore, many of the old preserving techniques cannot be rushed and are often dependent of external factors such as sun, wind, salt, and microorganisms that are all unaffected by advances in techniques and new tools.

◻ Drying *daikon* in Japan for making *tsukemono*, for example, traditional *takuan-zuke*.

The volume on preserving and fermenting came about after the author had spent a year travelling around in Japan to visit local food suppliers who grew rice, vegetables and fruit, produced soy sauce, *miso*, sake, rice vinegar, and fish sauce, and who had mastered all the traditional conservation techniques, including those for making *tsukemono*. But, as Nancy Hachisu writes, for her it was not enough only to visit these places, meet the artisans, and gather information that she could use in lectures, television presentations, articles, and books. It was necessary for her actually to prepare these ingredients with her own hands in order to be true to the food culture and take ownership of it. Only in this way could she gain a real understanding of the processes, giving her a sense of

self-confidence and a feel for her own abilities. The resulting attitude of "You can do it!" is deeply embedded in
Nancy Hachisu's work and reflected in her two wonderful,
very beautiful books.

Pickled Foods Made in Factories, Both Small and Large

The Gunma Prefecture, which lies north of Tokyo and can
be reached in an hour by the bullet train, is an agricultural
area that is renowned for its *tsukemono*, including large-
scale manufacturing plants. One day near the beginning of
June in 2016 one of us, Ole, had the opportunity to go to
Takasaki to visit both a small and a very large pickle factory. The trip was arranged by a Japanese colleague, Dr.
Kumiko Ninomiya, who is an internationally recognized
expert on umami. She had pulled some strings with her
business contacts and organized the tour of the establishments. Howie Velie, a professional chef and Director of
Education at the Culinary Institute of America, who takes
an interest in *tsukemono*, came along on the journey.
Kumiko-san acted as our guide and translator.

A Visit to a Typical Family Enterprise

At the station in Takasaki we were met by Kenzaburo
Takahashi, head of the public relations division of
Ajinomoto, the large, international corporation, who has
a network of contacts in the *tsukemono* factories. After a
lengthy journey by car through the very industrial area
around Takasaki, we arrived at Shitara, a modest family-
owned facility, which employs forty workers.

Hideo Shitara, the current director of this forty-year-
old firm, is a member of second generation of the family.
He and his wife greeted us outside the factory, surrounded
by stacks of boxes with fresh vegetables and fruits, waiting
to be turned into pickles.

Our first stop was a small meeting room where Mr.
Shitara recounted the history of the firm and told us about
their thirty or so different types of *tsukemono*. The factory
concentrates almost exclusively on *asa-zuke*, which are
made very rapidly and, consequently, must be sold almost

immediately. When in season, most of the vegetables are sourced from the Gunma Prefecture, although some ingredients are also brought in from other regions in order to keep production going all year round.

Our Japanese host, Kumiko-san, translated and it gradually dawned on us that we were the first foreigners who had ever visited the factory. Various types of pickles were laid out on a table for us to sample—*daikon*, Chinese cabbage (*hakuzai*), Japanese cucumbers (*kyuri*), turnip greens (*nozawana*), yellow onions (*tamanegi*), and Japanese plums (*ume*). All were made using the *asa-zuke* method that, in this case, is based on the use of a light brine and does not involve either dehydration, fermentation, or the use of preservatives. The *daikon* are processed two to three days after they are harvested. They are cleaned and placed for up to a week at a temperature of 5 °Celsius in brine in a large vat. At the end of this time, salt makes up about 4 percent of their contents and they have been reduced in weight by about 40 percent, because the salt has drawn out a great deal of water. Sea salt with a high calcium content is used, with the result that the finished *tsukemono* are very crisp. After the brining, the *daikon* are put into a 65 percent sugar solution for about a week. This solution is used over and over. The process takes about two weeks and the pickles are then placed in pouches with a variety of brines and seasonings, such as soy sauce, to add interesting tastes.

◘ Industrial production of *tsukemono* from *daikon*.

Next we were treated to some taste tests. We started with the *daikon*, which is prepared simply with brine and sugar. We dipped them in three different types of soy sauce, ranging from the sweet (*ama-kuchi*), to the sour (*usu-kuchi*), and to the ordinary, *koi-kuchi*, as well as in a mixture of soy sauce and plum juice (*ume*). These distinct tastes have their adherents in different parts of Japan—*ama-kuchi* in Kyushu and *usu-kuchi* and *koi-kuchi* in Kansai, the area around Kyoto and Osaka. In the Kanto region, which includes greater Tokyo, and in Tohoku, further north, the mixture with the plum juice is preferred. We moved on to sampling the cucumbers (*kyuri asa-zuke*) and the yellow onions (*tamanegi*), which turned out to have a surprisingly mild taste. The Chinese cabbage (*hakusai-zuke*) was seasoned with chili peppers and the seaweed *konbu*, which contributes umami.

At this point, while we were tasting the pickles, we felt the table shaking and rattling sounds coming from the cupboards along the wall. Kumiko-san said laconically that this was just a '2.' It occurred to us that we had just experienced one of the rather frequent minor earth tremors that are common in this part of Japan. The locals generally carry on as if nothing has happened.

To conclude the tasting, we tried a seasonal specialty, which is available for only about one month of the year—brined, small green Japanese plums (*ume*) of the variety that when ripe is used to make *umeboshi*. These little gems have just about the crispest mouthfeel imaginable and they almost burst when one bites into them. But they have a large stone, so caution is advisable. It is difficult to imagine that unripe fruit could have such a perfect taste and a sublime mouthfeel. This type of *asa-zuke* is eaten in Japan as a snack, accompanied by a cup of green tea. Kumiko-san commented that we were lucky to visit at the right time of year.

About 10,000 *daikon* are processed by the factory every day of the year. We started the factory tour in the large arrival area where *daikon*, Chinese cabbages, and onions are prepared for brining. The impressive, very uniform *daikon* we saw that day were grown by farmers in the Gunma area. They are sorted by hand into three size categories and a part or all of the tops are cut off. The cabbages are split open lengthwise. Onions are stripped of their outer skins using air pressure. Each type of vegetable is then placed in separate mild brines.

Going further into the factory, we went through a cold room that contained enormous vats with *daikon*, aubergines, cucumbers, and Chinese cabbages in a very strong salt solution. A large concrete block was placed on top of a lid on each vat to keep the vegetables completely immersed in the brine. They are left for several months and are made into a totally different type of pickle, *furuzuke*, which has a very long shelf life.

We proceeded on to the last area of the factory where the different kinds of *tsukemono* are cut into an assortment of sizes. They are then placed on a conveyor belt, sealed into plastic pouches with various marinades, and packaged, ready for shipping.

Tsukemono in a Large Factory Setting

After our visit to Shintara we were in for a study in contrasts as we moved on to one of Japan's biggest *tsukemono* factories, Shin-shin, in Maebashi, quite close to Takasaki. This is also a family-run company, now in the hands of the fourth generation, with a fifth one waiting in the wings to ensure the succession. The firm was established 122 years ago, employs 230 workers, and has a total annual output of about 10,000 tons of *tsukemono*, encompassing 100 different varieties.

Shin-shin is located next to one of Japan's largest and most important rivers, the Tonegawa, in order to have access to an abundant source of water. We were received in the conference centre by a large contingent of directors and communications staff. The head of the firm, Masami Kabayashi, described the company and its products. Shin-shin is actually a conglomerate of three food factories in Gunma, as well as two more in Japan and one in China. The company logo consists of three discs depicting green fields, a blue sky, and a blue sea, respectively. According to Mr. Kabayashi, they reflect the purity of Nature in its three elements. The logo also indicates that Shin-shin manufactures not only *tsukemono*, but also proteins, amino acids, vegetable pastes, starch, and other potato products.

◻ Industrial production of *takuan-zuke.*

The pickles produced at Shin-shin are mostly *furu-zuke*, which are prepared over an extended period of time, but also have a long shelf life. They are typically made with *daikon*, cucumbers, and aubergines. As it is not possible to obtain a sufficient supply of raw ingredients from within Japan, some, for example, *daikon*, are imported from China. The vegetables are first placed in a 20 percent brine, which over the course of several weeks reduces their water content by up to 65 percent. After undergoing this treatment, the vegetables look rather pathetic. They are cut into appropriately sized pieces and immersed in a series of water baths to remove part of the salt content. This step is of vital importance in determining the quality of the finished product. The pickles are next placed in plastic bags with a variety of marinades, which are then sealed and pasteurized. This ensures that they will keep for a long time, even without refrigeration.

The major product line at Shin-shin is *fukujin-zuke*, made from pieces of *daikon*, cut-up sword beans, and a selection of spices. It is a popular accompaniment for rice dishes with curry sauce. Another one of their standard pickles is *shiba-zuke*, a combination of cucumbers and aubergines in a marinade infused with red *shiso*.

Tsukemono at the Market and in Shops

Most Japanese people buy their *tsukemono* at the supermarket. The larger ones have one or more basement food sections with an incredible selection of high quality goods that one is unlikely to find anywhere else in the world. Here it is possible to stock up on *tsukemono* to one's heart's content, as many stores have at least a hundred different types. Apart from the ones sold in packages, it is possible to buy the pickles at small specialty stalls where one can sample and buy them fresh from the barrels. Those with the greatest selection carry both a seasonal assortment and *tsukemono* brought in from other parts of the country.

Sometimes one is lucky and finds a supermarket with a special food section located in the top story. This typically consists of an area where one can purchase very high quality foods that are beautifully packaged and suitable for presentation as gifts. In addition, there may be a section set up like an old-fashioned market with booths separated by a criss-cross pattern of alleyways. When one steps into such a setting one is immediately surrounded by loud sounds. The dialogue between the customers and the vendors touting their wares draws one in to explore further in order to see, taste, and purchase *tsukemono*.

For many Japanese there is a certain nostalgia associated with visiting such a marketplace. There are long queues by the vendors who have an assortment of seasonal pickles on offer. Here one can have a good look at the barrels of *nuka-zuke* and *nara-zuke*, *nara-zuke* which give off their characteristic aromas and have 'the smell of home.'

Old-Fashioned *Tsukemono* Shops

In many Japanese towns one can still find old, established shops that sell only *tsukemono*, often made in-house. On a side road in the centre of Kyoto there is a well-hidden shopping area where Murakami-ju, an establishment that is over 150 years old, is located. The owner is called Murakami and *ju* stands for the number 10, which is represented by a sign that displays a seal in the shape of a circle with a cross inside it. This same symbol decorates the banner that hangs over the entrance. Paving stones lead into the shop and cover the rustic floor inside. The coolness of the stones and the elegant wood finish on the chilled display cases immediately give one the impression that the products sold here are of the finest quality. Very few tourists ever find their way here and it is obvious that the customers are locals who come for the express purpose of buying their favourite *tsukemono* when in season. By far the majority of these are fresh *asa-zuke*, which will keep for only a few days. At the back of the store, there is a little counter where one can make arrangements with the staff to place orders for delivery.

◘ Murakami-ju, a 150-year-old *tsukemono* shop in Kyoto.

Tsukemono at a Street Market

Kyoto's famous Nishiki market is not far from Murakami-ju. The market can be considered as a sort of neighbourhood of its own, with streets running up and down and sideways, organized in such a fashion that each of them typically has shops that sell similar products. There are streets where one finds only clothing, food products, jewellery, shoes, and so on. Around the market itself there is a wealth of restaurants and specialty stores, including some that are some devoted entirely to *tsukemono*. These are easy to identify as large, traditional pickle barrels are often found at their entrance and the smells of *nuka-doko, miso, sake-kazu*, and *yukari* are unmistakable. All kinds of pickled vegetables—*daikon*, radishes, turnips, cucumbers, aubergines, cabbages—and *umeboshi* are heaped up in the barrels or openly displayed in their fermentation medium. They look very inviting and one is allowed to sample them before deciding on which ones to take home.

Tsukemono, Nutrition, and Wellness

In Asia, and particularly in Japan, *tsukemono* are considered healthy foods because, apart from their content of macronutrients like proteins and carbohydrates, they are a source of important minerals, vitamins, and antioxidants, as well as soluble and insoluble dietary fibres. Because these pickles, whether or not they have been fermented, are produced without heating, many of their nutrients, vitamins, and active enzymes are better preserved than those in vegetables that have been cooked. A possible drawback, however, may be that some of these same desirable components, such as proteins, are less bioavailable as they pass through the digestive tract than those in cooked vegetables.

The vegetables in *tsukemono* are variously affected by salt, sugar, and vinegar and, possibly, by being dehydrated and fermented. The balance achieved by careful preparation is, nevertheless, generally such that it is possible both to preserve a significant proportion of vital nutrients and to transform the original plant material into something that is easy to digest. A comparison of the treated vegetables with their raw counterparts indicates that the various techniques employed for making *tsukemono* succeed in reducing the amount and the activity of water in the finished products. As a consequence, they have a greater concentration of dietary fibre, nutrients, minerals, and vitamins. For example, dietary fibre, which is essential for the proper functioning of the digestive system and maintaining the flora (the human microbiota) in the intestines, for regulating sugar levels in blood, and for the excretion of carcinogens and other undesirable substances in the food, is typically three times as concentrated in *tsukemono*.

Other factors that enhance wellness also come into play. The various vegetables used to prepare *tsukemono* each contain a long list of bioactive substances, some of which are known to have beneficial properties. Antioxidants and vitamins are two examples. And doubtless, the delicious crunchiness of the pickles is an important factor in stimulating appetite and releasing saliva, which improves the mastication process, the first and most important step in promoting good digestion.

If one prepares *tsukemono* at home, it is essential to maintain a high standard of hygiene to prevent the growth of unwanted microorganisms. These can cause pickled foods to acquire unpleasant tastes and smells and, in the worst instances, can result in toxic substances.

Slightly Sour, a Little Tart

Acids of various kinds are present to some extent in most types of *tsukemono*. Citric and malic acids are especially common as they occur naturally in raw vegetables. In addition, acetic acid might be present from vinegar used in the marinade, and lactic acid is sometimes formed during fermentation. A certain level of acid in foods can increase appetite and stimulate the production of saliva and digestive juices, both of which help to burn off calories and fats. Acids are also largely responsible for the sour and slightly tart taste of *tsukemono*.

Together with salt, the acid content determines which microorganisms will be able to thrive in the pickles. Acids can neutralize pathogenic bacteria and, together with the salt, give rise to good growing conditions for useful bacteria such as the different lactic acid bacteria. They also have a great deal of influence on enzymatic activity, often slowing it down and allowing nutrients to be released over a longer time span, which can also be beneficial for the digestive process.

▣ *Takana-zuke:* Japanese mustard leaf *tsukemono*.

Vitamin Content

The vitamin contents of vegetables and their bioavailability depend largely on the specific vegetables in question and how they are handled and prepared. Storing them for an extended period of time and subjecting them to heat will rob them of many of their vitamins. But as these processes may also increase the extent to which vitamins are absorbed during digestion, it is difficult to make a precise statement about the effects of storage and heating. In the meanwhile, dehydrating, pickling, and salting, to say nothing of fermenting, will greatly facilitate release of the nutrients, including vitamins, from the bulk of the plant matter in the vegetables, making them more bioavailable.

The vegetables in *tsukemono* benefit from not having been subjected to heat at any point. In addition, the loss of water-soluble vitamins will be appreciably lower in the types of *tsukemono* made from vegetables that have been dehydrated before being marinated. The vitamin loss is even lower when the raw ingredients are dried at cooler temperatures.

Tsukemono are a good source of vitamins A, B, and C. Those with a short preparation time are rich both in vitamins A and C. Even though fresh vegetables and fruit might contain an abundance of vitamin C, some of it is gradually lost when the vegetables are marinated for extended periods of time. *Kasu-zuke* contain vitamins B_1 and B_3 that seep into the vegetables from the sake lees and are found in the small amounts of marinade that cling to the vegetables. *Nuka-zuke* take up vitamins B_1 and B_3 from the rice bran. The content of vitamin B_1 can be up to twelve times as great in preserved vegetables as that found in the raw ingredients. In earlier times the vitamin B content in *kasu-zuke* and *nuka-zuke* was an important element in the diet of the Japanese people as a complement to their rather limited food repertoire, which contained a great deal of polished white rice.

Desirable Bacteria, Fungi, and Enzymes

The pickling bed made from rice bran, *nuka-doko*, draws liquid out of the vegetables and gives rise to conditions that favour the growth of lactic acid bacteria (*Lactobacillicus*). Both *kasu-zuke* and *nuka-zuke* absorb lactic acid bacteria from their respective fermentation media. These bacteria enhance the digestive process in the stomach and the intestines, where they have a positive impact on the regulation of the intestinal flora. The healthful benefits are similar to those that are derived from eating yogurt and other cultured milk products. The famous claim made in 1900 by Nobel laureate Elie Metchnikoff that this leads to a longer life span is, however, still unproven. Nevertheless, there is general agreement that the probiotic bacteria, among them lactic acid bacteria, have many properties that promote good health.

Because *tsukemono* are not treated with heat, the natural enzymes found in the vegetables are unaffected and those that are produced in the course of fermentation remain intact. This allows them to help in breaking down the plant matter to more bioavailable nutrients. It is worth noting, however, that as soon as food is ingested these enzymes are, in turn, rapidly degraded by other enzymes in the digestive system.

Beneficial Effects of Fermentation

Many claims are made about the health-enhancing properties of fermented and cultured foods and of the microorganisms that are used to prepare them. They have been linked to combatting a variety of diseases, including cancer, diabetes, and autism. But few of these assertions are supported by scientific evidence.

There are, however, three specific outcomes that are worth noting. For example, many people have said that they feel better with respect to their digestion when they eat fermented foods. In addition, it is well known that fermentation leads to the formation and release of minerals, vitamins, and acids that are thought to promote better overall health. Other substances formed during fermentation can improve the taste of the food, having a beneficial effect by enhancing quality of life, as will be discussed later. Finally, some microorganisms are known to have inherent positive, probiotic properties.

Fermentation Can Facilitate the Release of More Readily Bioavailable Nutrients

There are many microorganisms and enzymes in our bodies that break down food into more readily bioavailable substances, for example, fats, sugars, amino acids, and unbound minerals. As fermentation concentrates the nutritional value of a food, it can be seen as a process that decreases the burden on the digestive system and increases the uptake of nutrients—in a sense, 'pre-digesting' the food. In some cases, this involves enzymes and microorganisms that are able to do what our own digestive system cannot, that is, single out and release specific nutrients. Some of these processes can also break down toxins, for example, in cassava, so that the root becomes suitable for human consumption. In addition, fermentation can change the texture of certain foods, such as the tough parts of plants, leaving them softer and easier to chew.

Fermentation Helps to Preserve Foods So That They Are Safe to Eat and Will Keep Longer

It goes without saying that it is better and safer to eat fermented foods than ones that have spoiled. In earliest times, fermentation was probably the most effective method of conserving foods. This is how we came to have foods such as cheese, yogurt, soy sauce, air-dried hams, beer, and wine. These fermented products are now so firmly entrenched in many food cultures that we pay scant attention to just how many of them we eat and drink every day.

Fermentation Can Improve the Taste of Foods and Their Ability to Stimulate the Appetite and Regulate Food Intake

Fermented foods often have an unpleasant smell. For example, a ripe, old cheese or fermented fish sauce could be considered 'stinky' but, nevertheless, be palatable or at least have a taste to which we have become accustomed. An illustration of how this relates to wellness is that fermentation can lead to the development of certain substances that impart umami. This involves both the breakdown of proteins into free amino acids and the degradation of nucleic acids to free nucleotides. In the course of evolution, people everywhere have been drawn to umami as a positive taste sensation, precisely because it is an indication of bioavailable substances in the form of proteins and amino acids.

Research done in the past few years has shown that umami stimulates the appetite and increases saliva flow, which indirectly serves to strengthen the immune system, as well as facilitating mastication. All of these factors contribute to well-being. Another recent discovery is that there are dedicated umami receptors throughout the gastro-intestinal system and a special communications axis between the brain and the stomach. The brain sends signals along this route to tell the stomach to release digestive enzymes—proteins are on the way! Later a message goes in the opposite direction to convey feelings of fullness and satisfaction from the stomach to the brain. As a result, the umami tastants that are formed during fermentation not only contribute to deliciousness and stimulate the appetite, but also couple into a homeostatic mechanism that is of great importance in regulating food intake. Both aspects of the effects of umami promote wellbeing even though, paradoxically, they work in opposite ways. The homeostatic mechanism is useful for those of us who have a tendency to overeat, while stimulation of the appetite helps those who have lost interest in food because they are sick or elderly and, consequently, become malnourished or weigh too little.

◘ Japanese sea salt, *moshio*, that has been produced by evaporation of seawater with some seaweed ashes.

Go Easy on the Salt

Most commercially prepared food products are seasoned with fats, sugar, and above all with salt in order to improve their taste. In many countries, the amount of salt added has increased in the last few years and in several of them it makes up about 80 percent of the normal daily intake. The World Health Organization has estimated that the average daily salt intake of 9–12 grams is twice as big as the recommended 5 grams, which is equivalent to about a teaspoonful. The members of the organization have resolved to decrease global consumption of salt by 30 percent by 2025.

The underlying reason for the decision to reduce salt consumption can be found in the large-scale epidemiological studies that indicate that the world-wide increase in cardiovascular disease and high blood pressure can to a certain extent be blamed on excessively salty food. The relationship between the salt content and the prevalence of these diseases has, however, turned out to be more complex than initially anticipated. The over-use of salt ($NaCl$) is especially problematic because it has been accompanied by a reduction in the consumption of fruits, vegetables and other foods that have an abundance of dietary fibre. Fresh produce, as well as seaweeds, are significant sources of potassium salts, which are known to lower blood pressure.

It is likely that the growing preference on the part of younger Japanese consumers for the types of *tsukemono* that are less salty and have a shorter shelf life, for example, *asa-zuke*, is linked to concern about the potentially harmful effects of a high salt intake. In the course of the past twenty years, the amount of salt in popular Japanese pickles made from *daikon* and Chinese cabbage, among others, has been reduced by a half.

To a certain extent, preparing *tsukemono* using vinegar and an alcohol such as sake can compensate for a lower salt content, without compromising the taste. And the appropriate use of preserving agents and cooling techniques can extend the keeping qualities of these lightly salted pickles.

◘ Tasting samples of *tsukemono* at a food market in Japan.

Eat *Tsukemono* in Moderation

In order to derive the most benefit from eating *tsuke-mono* as part of the daily diet, it is best to do so keeping in mind that it was originally intended to be an appetite stimulant and an accompaniment to the main dishes. As many of them are rich in umami, they can make a contribution to the taste of other vegetables that lack this basic taste, often without the necessity to add fats. In addition, because umami interacts synergistically to enhance salty tastes, a small serving of *tsukemono* can make it possible to reduce the salt content of the other foods.

There are possible drawbacks to eating large quantities of salt-pickled and fermented vegetables. As discussed above, one is related to their salt content. The other is that the they may contain carcinogenic substances that are formed in conjunction with the taste enhancing substances by the action of the microorganisms involved in the fermentation process. But there seem to be fewer grounds for concern about the acidity of these products, even though a number of researchers have pointed out that some of the nutrients in fresh vegetables are degraded in this type of environment.

Many extensive investigations of the potentially harmful effects of pickled and fermented vegetables have been carried out, in particular among population groups in Asia, including China, South Korea, and Japan, where these are eaten in large quantities. While the results are not conclusive, there appears to be a relationship between the intake of these products and the incidence of stomach cancer. This effect can be counteracted by consuming a significant quantity of vegetables, which in the case of Koreans and Japanese is three times and twice as great, respectively, as that of many Europeans and North Americans. The abundance of dietary fibre and antioxidants in fresh produce are well-known positive factors in decreasing the risk of various types of cancer.

Comparative studies seem to indicate that fermented vegetables can cause cancer of the esophagus. In this case also, there are meaningful differences between the various studies and it is not possible to establish a clear correlation between the possible presence of carcinogenic substances in fermented vegetables and this quite rare type of cancer. In some instances these substances, the so-called nitroso compounds, are secreted by fungi that grow on the vegetables due to inadequate control of how the finished products are prepared and stored.

Under all circumstances, it is wise to ensure that only a lesser proportion of one's vegetable intake is in the form of pickled and fermented vegetables. *Tsukemono* should, therefore, be eaten in moderation, especially those that have a high salt content.

Wabi, Tsukemono, and Esthetics

No description of *tsukemono* is complete without considering its relationship to *wabi*, an esthetic concept that has religious roots and is often associated with Zen Buddhism. There is no direct translation of *wabi* and one depends on understanding its meaning through its connotations of minimalism, freshness, and understated elegance. *Wabi* is often described as complex esthetic concept that is used to characterize a person, a thing, or a form of life that exhibits modesty, humility, loneliness, sadness, simplicity, or stillness.

As described in this book, the preparation of *tsukemono* is closely linked to the idea of *wabi* and is built on a foundation of simplicity and respect for preserving the appearance, taste, and mouthfeel of the vegetables and fruits. And, to a certain extent, their various shapes and colours are also integral features that create an interesting visual appeal and add an element of complexity to presenting them in an esthetic manner.

According to the traditions of Japanese food culture, a meal should incorporate five different colours: red, green, yellow, white, and black. *Tsukemono*, on their own, can help to achieve this goal by contributing red *beni-shoga*, green *kyuri asa-zuke*, yellow *daikon nuka-zuke*, white *senmai-zuke*, and *kasu-zuke* that are made from aubergines that have been fermented over such a long period of time that they are almost black.

In the classical Japanese cuisines of *washoku* (food in the traditional, indigenous style), *kansha* (vegan and vegetarian cooking), and *kaiseki* (elaborate, multi-course meals) all of these aspects help to create harmony among the different sensory impressions associated with the meal.

Tsukemono are meant to be eaten in small quantities selected from a variety of different types. These tasty, nutritious condiments were, in the first instance, a simple staple that was used to round out an ordinary meal of plain boiled white rice and *miso* soup. But despite their humble origins they were also considered sufficiently sophisticated to be integrated into temple foods and very elaborate dinners. Virtually no meal in Japan is now served without at least a few pickles and even in the least traditional households some small pieces usually find their way on to the plate.

The way in which *tsukemono* are presented is rather Zen-like—it is a question of *wabi*. The pickles are usually served separately in a small bowl or on a plate that is carefully selected to complement the arrangement. The structure of the vegetables and the difference between their external and internal appearances can be accentuated by slicing them both lengthwise and crosswise. They are cut into various shapes, no bigger than a single bite that can be picked up with chopsticks or the fingers. They are not piled up but placed so that each individual piece is allowed to stand out modestly on its own. The concept is based on placing as much emphasis on the empty spaces between them as on the actual spaces they fill.

In her wonderful book, *Kansha,* about Japanese vegetarian cuisine, Elizabeth Andoh writes about the presentation of finely sliced, salted vegetables and *tsukemono* in the style called *ten mori,* which means 'heavenly arrangement.' The vegetables are pressed together in the palms of the hands to form a cylinder or pyramid that is then made to stand on a plate so that it points heavenward. In this way, a small quantity of pickled ingredients can be made to look both simple and beautiful and appear to be much more than they actually are.

◘ Simple arrangement of cucumber *tsukemono.*

Tsukemono should not be thought of as foods that are eaten only in Japan or that belong exclusively to its culinary culture. Rather, these pickles represent a concept, a principle of how to eat, that is just as applicable in other cuisines and that can be a source of inspiration. As shown in some of the recipes in this book, it is very easy to incorporate *tsukemono* into other dishes such as salads, not least to add an interesting, crunchy mouthfeel.

The authors of this book are convinced that armed with knowledge about *tsukemono* and the way in which they are prepared, we can find pleasure in discovering a treasure trove of new ways to enjoy eating vegetables. In addition, many of them are easy to make at home, keep well in the refrigerator for extended periods of time, and consequently can be ready at a moment's notice to add a little spark to any meal. This can lead to more esthetically presented dishes that, as a bonus, are nutritious and healthy.

◘ Smoking of black turnips for *tsukemono*.

The Technical Details

Glossary of Japanese Terms

agehama an old, very labour-intensive technique for producing salt from seawater. The seawater is repeatedly poured over sandy terraces on the beach to evaporate some of the water and concentrate the salt content. The resulting briny liquid is then heated in special drying ovens to extract the salt.

aka-miso red *miso* made with rice.

amazu-zuke *tsukemono* marinated in vinegar with sugar added to it.

amazake ('sweet sake') sweet, non-alcoholic drink, made by fermenting rice with rice *koji*; sometimes flavoured with ginger. *Amazake* can be used to make *koji-zuke*.

ao-nori dried seaweed or seaweed sheets produced from green marine algae, for example, *Monostroma* or *Ulva*.

asa-zuke ('shallow pickling') *tsukemono* prepared over a very short period of time, anywhere from thirty minutes (for *sokuseki-zuke*) to overnight (*ichiya-zuke*). The most quickly and easily made varieties (*shio-zuke*) are prepared simply by rubbing the vegetables with salt and placing them in a sealed bag.

beni-shoga pickled, fresh ginger shoots or sliced ginger root that are first salted and then marinated in vinegar; often in plum vinegar (*umezu*).

bento meal served in a box that is divided into compartments.

bettara-zuke classical type of *koji-zuke* from the Tokyo area made from *daikon* that is placed in a marinade of sugar, salt, and sake together with *koji*. It has a sticky and moist mouthfeel.

daikon large, white Chinese radish (*Raphanus sativus*).

dashi ('boiled extract') broth based on an extract of a seaweed (*konbu*) and bonito fish flakes (*katsuobushi*). First *dashi* (*ichiban dashi*) and second *dashi* (*niban dashi*) refer to the first and second extracts, respectively. *Konbu dashi* is based entirely on *konbu*. *Shojin dashi* is a purely vegetarian *dashi*, in which mushrooms (*shiitake*) replace the *katsuobushi*. *Dashi* is the principal source of umami in Japanese cuisine.

doko fermentation medium made from bran, for example, rice bran (*nuka-doko*) or wheat bran.

fu Japanese expression for wheat gluten, also known in Chinese as *seitan*; in raw form it is *nama fu* and roasted or dried it is known as *yaki fu*.

fukujin-zuke the best known and most common type of *shoyu-zuke*, made from *daikon*, cucumbers, aubergines, sword beans, and lotus roots. It can be served as a relish or chutney with rice and curry dishes.

furu-zuke *tsukemono* that, in contrast to *asa-zuke*, are prepared over a long period of time, often involving fermentation, for example, *nuka-zuke* and *nara-zuke*.

gari sushi-bar slang for sweet and sour pickled ginger root, a type of *amazu-zuke*; often sliced very thinly and served with sushi and coloured with red *shiso*. The Japanese name for ginger root is *shoga*.

haiku short Japanese poem traditionally written in seventeen syllables.

hakusai-zuke *shio-zuke* made from Chinese cabbage (*Brassica rapa*), often mixed with carrots and cucumber and seasoned with *konbu*, *togorashi*-chili, and *yuzu* peel. *Hakusai-zuke* are eaten as a salad or as a condiment, for example, with fish dishes.

hanami the Japanese custom of taking the time to enjoy the fleeting beauty of the cherry blossoms in the spring.

hashi-yasume ('vacation for the chopsticks') denotes a time when the chopsticks are given a rest while one eats a very small portion of *tsukemono*.

iburi-gakko speciality from the Akita Prefecture; made from *daikon* which is smoked, often using cherry wood, and then immersed in *nuka-doko*.

ichiya-zuke pickles that are allowed to cure overnight, also referred to as *asa-zuke.*

ita-zuri age-old Japanese technique in which vegetables are rubbed thoroughly with salt before they are pickled. This helps to soften the vegetables and tenderize the skin, and it increases the amount of water that can be drawn out of them.

kabocha Japanese pumpkin (*Cucurbita moschata*).

kabu Japanese turnip (*Brassica campestris* var. *glabra*).

kaiseki formal Japanese meal, in which the dishes are presented in a prescribed order. Great emphasis is placed on serving fresh produce that is in season. *Cha-kaiseki* is the formal meal that is served before a Japanese tea ceremony.

kansha Japanese expression that encompasses placing value on something. It is used in connection with food, especially vegetables, as a way to acknowledge Nature's gifts and human ability to convert these into delicious meals.

karashi strong Japanese mustard (*Brassica nigra, Brassica juncea, Sinapis alba*).

karashi-zuke *tsukemono* made with *karashi*.

kasu lees.

kasu-zuke *tsukemono* marinated in *kasu*.

katsuobushi cooked, salted, dried, smoked, and fermented *katsuo* (bonito), which is shaved into paper-thin flakes, used, for example, to prepare *dashi*.

koji fermentation medium made from rice, barley, or soy bean paste that is seeded with spores from the fungus *Aspergillus oryzae*. *Koji-kin* is a freeze-dried product, which is a mixture of spores and rice flour and that can be used as a starter for a *koji* culture.

koji-zuke *tsukemono* prepared using the special fermentation medium *koji*, which, among other ingredients, contains microscopic fungi (*Aspergillus oryzae*).

komesu traditional Japanese rice vinegar made from polished brown rice.

kurosu traditional Japanese rice vinegar made from unpolished black rice.

konbu (*kombu*) species of large brown alga (*Saccharina japonica*). Among other uses, it is the essential ingredient in *dashi*. *Konbu* contains large quantities of free glutamate and is a source of umami.

konbu-tsukudani *konbu* simmered in *mirin* and soy sauce.

konomono ('pickled things with a pleasant aroma') older expression for *tsukemono*.

kyuri Japanese cucumber (*Cucumis sativus*).

kyuri asa-zuke quickly pickled *asa-zuke* made with Japanese cucumbers.

maki-zushi sushi roll, with a sheet of *nori* wrapped either inside it or around it.

mirin sweet rice wine with ca. 14 percent alcohol content; made from cooked rice that is seeded with the fungus *Aspergillus oryzae*. The rice is mixed with *shochu*, distilled rice brandy. The enzymes in the fungus break down the starch in the rice to sugar and its proteins to free amino acids, among them glutamate, which impart umami. *Mirin* is not drunk on its own but is used exclusively to season food.

miso paste made from soybeans, sometimes mixed with a cereal grain, that is fermented with the help of salt and *koji*. It typically contains 14 percent protein and great quantities of free amino acids, especially glutamate. The salt content varies from 5–15 percent. Common types are white *miso* (*shiro-miso*), red *miso* (*aka-miso*), yellow or light brown *miso* (*shinshu-miso*), and barley *miso* (*mugo-miso*). *Miso* soup is a light Japanese soup made with *dashi*, to which *miso* is added.

miso-doko *miso* fermentation medium, used to make *miso-zuke*.

miso-zuke *tsukemono* made with vegetables pickled in *miso*, for example, *aka-miso*. *Nasu-miso* are *miso-zuke* made with aubergines.

moromi thick paste of grains and soybeans that undergoes slow fermentation as one stage of the process that is used to produce Japanese soy sauce, *shoyu*.

moshio traditional Japanese sea salt produced by boiling and reducing seawater to which seaweed ash has been added.

mottainai Buddhist expression that refers to feelings of regret and sadness caused by having allowed something to go to waste or be misused.

nara-zuke one of the best known and tastiest kind of *kasu-zuke*. It is made from a variety of vegetables, but cucumbers, marrows, and Japanese pickling melons (*uri*) result in a good texture. The longer the ingredients are left in the marinade, the darker their colour. Typically, these pickles keep well, up to several years.

nara-zushi the oldest form of sushi, in which rice fermented over a long period of time acts as a preservative for the fish, for example, carp made into *funa-zushi*. The rice is not eaten but discarded.

nasu karashi-zuke *tsukemono* made from aubergines in a marinade with *karashi*.

nori fronds of the red alga *Porphyra*, which are dried, pressed, and roasted to make paper-thin sheets that are used, for example, to make *maki-zushi*.

nori-tsukudani *nori* (*Porphyra*) simmered in *mirin* and soy sauce.

nozawana the green stems and leaves of Japanese turnip greens (*Brassica campestris* var. *hakabura*).

nuka rice bran.

nuka-doko (*toku*) fermentation medium made with rice bran. One of its uses is for the preparation of *takuan-zuke* made with *daikon*.

nuka-miso ('smelly women') expression used to characterize those women whose hands take on a peculiar odour and yellow colour from using them every day to stir the *nuka-doko*, the rice bran fermentation medium used to make certain types of *tsukemono*.

nuka-zuke *tsukemono* prepared by placing vegetables in a rice bran fermentation medium (*nuka-doko*), for example, *takuan-zuke*.

okoko another name for *oshinko*.

oshinko (*shinko*) salt pickled vegetables, often lightly pickled, that are not completely preserved. *Oshinko* in the form of *takuan-zuke* are used to make *oshinko-maki*.

ponzu liquid seasoning composed of soy sauce, *dashi*, and *yuzu* juice, to which a little sake is sometimes added; can be used as a marinade.

rakkyo *su-zuke* made with Chinese onions (*Allium chinense*) or possibly spring onions or shallots.

ramen wheat flour noodles.

ryorishu cooking sake; contains less alcohol and may have a 2–3 percent salt content that makes it unfit for drinking.

sake rice wine, made by fermenting cooked, polished rice using the medium *koji*, which contains enzymes that can break down starch to sugar and proteins to free amino acids. A yeast culture then converts the sugar to alcohol. Cooking sake (*ryorishu*) has a lower alcohol content and is suitable only for making food.

sake-kasu sake lees.

sakura cherry blossoms. *Sakura no hana no shio-zuke* are salt pickled cherry blossoms.

sakura-mochi ball of sweet glutinous rice with a pickled cherry leaf wrapped around.

sansho type of Japanese pepper (*Zanthoxylum piperitum*), which is similar to Chinese Sichuan pepper.

sashimi sliced raw fish or shellfish.

sato-zuke a type of *tsukemono* which is distinguished from other kinds because no salt is involved in its preparation. *Sato-zuke* it is made by candying the ingredients in a sugar solution that is allowed to simmer for several days. *Sato-zuke* are similar to candied fruit. It is often made with Asian pickling melon, lotus roots, ginger root, *ume*, and *yuzu* peels.

senmai-zuke ('pickling in a thousand layers') speciality from Kyoto, made up of paper-thin slices of turnip (*kabu*) placed in a barrel together with *konbu* in a marinade of sweet rice vinegar.

shiba-zuke salt pickled aubergines and cucumbers, typically coloured and seasoned with red *shiso*.

Shinto ('way of the gods') traditional religion of Japan that emphasizes the divinity that is manifested in natural objects.

shio Japanese expression for sea salt.

shio-furi brining that occurs when salt is sprinkled on fresh vegetables, especially cut-up ones, allowing them to 'sweat' to extract water. They are then covered with water or a marinade.

shio-koji salted rice paste inoculated with *koji*. *Ikitai shio-koji* is a filtered variety of *shio-koji* that is clearer and flows more easily.

shio-momi brining by sprinkling salt onto especially thinly sliced vegetables, stirring them, and pressing the salt into them by hand.

shio-zuke ('shallow pickling') the easiest and most quickly made *tsukemono*, also called *asa-zuke*. The process takes as little as thirty minutes or as long as overnight.

shiro-miso white, sweet *miso* that is produced in Kyoto.

shiro-uri salt pickled *uri*, often marinated in *sake-kasu* (*shio-uri kasu-zuke*).

shiso a herb from the mint family (*Perilla frutecens*), that comes in three varieties, red (*aka-jiso*), green (*ao-jiso*), and green-red (*aoaka-jiso*).

shiitake mushroom (*Lentinus edodes*), which in dried form contains large quantities of guanylate, a source of synergistic umami.

shochu distilled rice brandy with an alcohol content of 36–45 percent.

shoga Japanese word for ginger root (*Zingiber officinale*).

shojin ryori classical, vegetarian temple food prepared in accordance with Buddhist precepts. It was first introduced into Japanese temples in the sixth century CE but was not widely known until the arrival of the Zen school of Buddhism about 700 years later. The meals consist principally of products based on soybeans, including tofu, *miso*, and *shoyu*, as well as fungi, seaweeds, and *fu*.

shokutaku tsukemono ki pickling barrel for making *tsukemono*; now generally refers to modern plastic crocks.

shoyu Japanese soy sauce.

shoyu-zuke *tsukemono* marinated in a mixture of soy sauce and sake or, possibly, *mirin* if the pickles are to be sweeter. An example of this type is *fukujin-zuke*.

soba buckwheat noodles.

su rice vinegar containing about 4.2 percent acetic acid.

sunomono ('things that allow themselves to be prepared with rice vinegar') these can be side dishes such as a little salad, with or without greens, in a tart marinade.

sushi Japanese food style made with cooked sweet, sour, and salty white rice combined, for example, with raw fish, shellfish, seaweed, vegetables, omelette, and mushrooms.

su-zuke *tsukemono* marinated in rice vinegar (*su*).

takuan-zuke *nuka-zuke* made with *daikon*, often coloured with turmeric so that it turns yellow. It is thought to have been named after a Zen monk, Takuan Soho (1573–1645).

tamanegi yellow onions (*Allium cepa*).

takana-zuke *tsukemono* of Japanese mustard leaf (*Brassica jucea* var. *integrifolia*).

taru traditional wooden barrel made from cedar or cypress wood, used for pickling and fermenting.

tataki preparation technique for raw fish. The filet is either chopped finely or seared quickly on all surfaces.

tawashi traditional kitchen scrub brush made from hemp fibre.

ten mori ('heavenly arrangement') arrangement of finely sliced seafood or *tsukemono* pressed together in the palms of the hands to form a cylinder or pyramid that is made to stand on a plate so that it seems to be pointing heavenward.

teriyaki grilled fish, chicken, and vegetables dipped in a sauce made from flour, *mirin*, sake, and sugar.

tofu coagulated protein-rich solid prepared from soy milk.

togorashi Japanese chili.

tsuke something that has been soaked or marinated. When it follows another word, *tsuke* is changed to *–zuke*.

tsukemono Japanese pickling technique. *Tsukemono* (つけもの, 漬物), pronounced 'skay-moh-noh' means 'something that has been soaked or marinated,' from *tsuke* meaning 'hydrated' and *mono* meaning 'thing.'

tsukudani a method for preparing and conserving vegetables, seaweed, fish, and shellfish. In contrast to *tsukemono*, which are prepared cold, it involves using heat to simmer the ingredients over a long period of time in soy sauce, *mirin*, and sugar until they are either dry or sticky. Examples made with seaweed include *konbu-tsukudani* and *nori-tsukudani*.

udon thick, soft wheat noodles.

umami ('delicious essence') Japanese expression for the fifth basic taste, a term suggested by the Japanese chemist Kikunae Ikeda in 1909 in connection with the identification of glutamate derived from the *konbu* in *dashi*. Umami has two components, a basal one from free glutamate, and a synergistic part due to the simultaneous presence of 5'-ribonucleotides especially inosinate found, for example, in *katsuobushi* and guanylate from *shiitake*.

ume small stone fruit (*Prunus mume*) that is like an apricot or a plum; cultivated in Japan. The wine made with it is sometimes referred to as 'plum wine.'

umeboshi expression for the most traditional type of salt pickled *tsukemono* (*shio-zuke*). The word combines the name of the fruit, *ume*, with *boshi* meaning 'dried.'

umezu vinegar made from *ume*.

uri Asian pickling melon (*Cucumis melo* var. *conomon*) that is very well suited for making *tsukemono*, for example, *shiro-uri kasu-zuke*, which is *uri* marinated in sake lees.

wabi complex esthetic concept that is used to characterize a person, thing, or form of life that exhibits modesty, humility, loneliness, sadness, simplicity, or stillness.

wakame brown seaweed species (*Undaria pinnatifada*); often incorporated into soups and marinades for preparing *tsukemono*.

washoku expression denoting harmony in the food, linked to traditional Japanese cuisine or meals.

wasabi Japanese horseradish (*Wasabia japonica*).

yukari salted, dried red *shiso*.

yuzu small Japanese citrus fruit (*Citrus junus*) with a more aromatic taste than a lemon.

zuke derived from the expression *tsuke*, which means 'soaked.' When placed after another word, *tsuke* changes to -*zuke*, for example, in *miso-zuke*, which are *tsukemono* prepared using *miso* as a fermentation medium.

Illustration Credits

All photographs except those listed below were taken by Jonas Drotner Mouritsen. p. x (bottom) (Charles Zuckermann); 19 (top), 21 (bottom); 27 (top), 82, 125, 129, 132, 134, 136 (Ole G. Mouritsen); pp. 53 (Mathias Porsmose Clausen); p. 127 (Hideo Shitara).

Bibliography

Andoh, E. *Washoku: Recipes from the Japanese Home Kitchen.* Ten Speed Press, Berkeley, 2005.

Andoh, E. *Kansha: Celebrating Japan's Vegan and Vegetarian Traditions.* Ten Speed Press, Berkeley, 2010.

Bibbins-Domingo, K., G. M. Chertow, P. G. Coxson, A. Moran, J. M. Lightwood, M. J. Pletcher & L. Goldman. Projected effect of dietary salt reductions on future cardiovascular disease. *New Engl. J. Med.* **362**, 590-599, 2010.

Bitterman, M. *Salted: A Manifesto on the World's Most Essential Mineral.* Ten Speed Press, Berkeley, 2010.

Bourne, M. *Food Texture and Viscosity.* Academic Press, London, 2002.

Chai, B. C., J. R. an der Voort, K. Grofelnik, H. G. Eliasdottir, I. Klöss, & F. J. A. Perez-Cueto. Which diet has the least environmental impact on our planet? A systematic review of vegan, vegetarian and omnivorous diets. *Sustainability* **11**:4110, 2019.

Felder, D., D. Burns & D. Chang. Defining microbial terroir: The use of native fungi for the study of traditional fermentative processes. *Int. J. Gast. Food Sci.* **1**, 64–69, 2012.

Fuji, M. *The Enlightened Kitchen: Fresh Vegetable Dishes from the Temples of Japan.* Kodansha Intnl., Tokyo, 2005.

Hachisu, N. S. *Japanese Farm Food.* Andrews McMeel Publishing, LLC, Kansas City, 2012.

Hachisu, N. S. *Preserving the Japanese Way: Traditions of Salting, Fermenting, and Pickling for the Modern Kitchen.* Andrews McMeel Publishing, LLC, Kansas City, 2015.

He, F. J. & G. A. MacGregor. Reducing population salt intake world-wide: from evidence to implementation. *Prog. Cardiovasc. Dis.* **42**, 363–382, 2010.

Hisamatsu, I. *Quick and Easy Tsukemono: Japanese Pickling Recipes.* Joie, Inc., Tokyo, 1999.

Hosking, R. *A Dictionary of Japanese Food: Ingredients and Culture.* Tuttle Publishing, Boston, 1996.

Inden, H., Y. Kawano, Y. Kodama & K. Nakamura. Present status of vegetable pickling in Japan. *Acta Horticulturae* **483**, 421–428, 1999.

Islami, F., J.-S. Ren, P. R. Taylor & F. Kamangar. Pickled vegetables and the risk of oesophageal cancer: a meta-analysis. *Brit. J. Cancer* **101**, 1641–1647, 2009.

Katz, S. E. *The Art of Fermentation.* Chelsea Green Publishing, White River Junction, Vermont, 2012.

Levenstein, H. *Fear of Food: A History of Why We Worry about What We Eat.* University of Chicago Press, Chicago 2012.

Makki, K., E. C. Deehan, J. Walter & F. Bäckhed. The impact of dietary fiber on gut microbiota in host health and disease. *Cell Host & Microbiota* **23**, 705–715, 2018.

McGee, H. *On Food and Cooking: The Science and Lore of the Kitchen.* Scribner, New York, 2004.

Mouritsen, O. G. *Sushi: Food for the Eye, the Body & the Soul.* Springer, New York, 2009.

Mouritsen, O. G. *Seaweeds: Edible, Available, and Sustainable.* The University of Chicago Press, Chicago, 2013.

Mouritsen, O.G. Deliciousness of food and a proper balance in fatty acid composition as means to improve human health and regulate food intake. *Flavour* **5**:1, 2016.

Mouritsen, O. G. *Tsukemono* - crunchy pickled foods from Japan: a case study of food design by gastrophysics and nature. *Int. J. Food Design* **3**, 103124, 2018.

Mouritsen, O. G. & K. Styrbæk. *Umami: Unlocking the Secrets of the Fifth Taste*. Columbia University Press, New York, 2014.

Mouritsen, O. G. & K. Styrbæk. *Mouthfeel: How Texture Makes Taste*. Columbia University Press, 2017.

Mouritsen O. G. & and K. Styrbæk. Design and 'umamification' of vegetables for sustainable eating. *Int. J. Food Design* **5**, 9–42, 2020.

Murooka, Y. & M. Yamshita. Traditional healthful fermented products of Japan. *J. Ind. Microbiol. Biotechnol.* **35**, 791–798, 2008.

Myhrvold, N. *Modernist Cuisine: The Art and Science of Cooking*. The Cooking Lab Publ., USA, 2010.

Nguyen, Q. V. *Pickled and Dried Asian Vegetables*. A Report for the Rural Industries Research and Development Corporation 00/**45**, Kingston, 2000.

O'Donnell, M., A. Mente, S. Rangarajan *et al.* (for the PURE Investigators). Urinary sodium and potassium excretion, mortality, and cardiovascular events. *New Engl. J. Med.* **371**, 612–623, 2014.

Ogawa, S. *Easy Japanese Pickling in Five Minutes to One Day*. Japan Publications Trading, Tokyo, 2003.

Ren, J. S., F. Kamangar, D. Forman & F. Islami. Pickled food and risk of gastric cancer – a systematic review and meta-analysis of English and Chinese literature. *Cancer Epidemiol. Biomarkers Prev.* **21**, 905–915, 2012.

Richie, D. *A Taste of Japan*. Kodansha, Tokyo, 1985.

Schatzker, M. *The Dorito Effect: The Surprising New Truth About Food and Flavour*. Simon & Schuster, New York, 2015.

Shih, J. & J. Umansky. *Koji Alchemy: Rediscovering the Magic of Mold-Based Fermentation*. Chelsey Green Publishing, London, 2020.

Shimizu, K. *Tsukemono. Japanese Pickled Vegetables*. Shufunotomo Co., Ltd., Tokyo, 1993.

Solomon, K. *Asian Pickles.* Ten Speed Press, Berkeley, 2014.

Springmann, M. *et al.* Options for keeping the food system within environmental limits. *Nature* **562**, 519–525, 2018.

Tsuji, S. *Japanese Cooking: A Simple Art*. Kodansha International, Tokyo, 1980.

Vinther, C. V. & O. G. Mouritsen. The solution to sustainable eating is not a one-way street. *Front. Psychol.* **11**:531, 2020.

Willett, W. *et al.* Food in the Anthropocene: the EAT-Lancet Commission on healthy diets from sustainable food systems. *Lancet* **393**, 447–492, 2019.

Welbaum, G. E. *Vegetable Production and Practices*. CABI, Oxfordshire, 2015.

Wrangham, R. *Catching Fire: How Cooking Made Us Human*. Basic Books, New York, 2009.

Yamaguchi, E. *The Well-Flavoured Vegetable*. Kodansha International, Tokyo, 1988.

Index